Handbook of
BASIC STATISTICAL CONCEPTS
For Scientists and Pharmacists

Handbook of
BASIC STATISTICAL CONCEPTS

For Scientists and Pharmacists

Shubha Rani

Alpha Science International Ltd.
Oxford, U.K.

**Handbook of
BASIC STATISTICAL CONCEPTS**
For Scientists and Pharmacists
242 pgs. | 50 figs. | 58 tbls.

Shubha Rani
Biometrics and Data Management
Synchron Research Services Pvt. Ltd.
Sarkhej Gandhinagar Highway
Ahmedabad

ALPHA SCIENCE INTERNATIONAL LTD.
7200 The Quorum, Oxford Business Park North
Garsington Road, Oxford OX4 2JZ, U.K.

www.alphasci.com

ISBN 978-1-84265-700-3

Printed in India

Dedicated to
the memories of
my parents Promod and Madhuri Khare
&
brother Rajiv Ranjan

"I myself do not burden my memory with simple facts that can be looked up in textbooks. The true purpose of education is to train the mind to think"

Albert Einstein

Preface

The idea of this book originated during my thirteen-year tenure in academia. Though statistics have been applied in pharmaceutical and medical sciences since long, I felt pharmacy students and researchers are still not aware of its basic concepts. They mostly apply statistical methods without actual conceptual understanding. In fact, a number of colleagues and friends say that they would like to learn about statistical methods, but find it difficult to do so from the existing literature. Also, I have come across very few books written targeting pharmacy students and scientists. This ignited the idea of writing a book on basic concepts of statistics with examples from pharmacy. My prime interest was to make them understand why statistical methods are applied in pharmaceutical sciences and the difference between various methods available in literature.

This book is designed following the view of David Hilbert (1862-1943) - "A mathematical theory is not to be considered complete until you have made it so clear that you can explain it to the first man whom you meet on the street". This book is an attempt to explain the basic concepts of statistics to pharmacy students and scientists and is written in very simple and understandable language with the objective of making them understand the basic statistical methods which are applied day in and day out in pharmaceutical research. The aim is to introduce students to understand the problem in hand appropriately and then apply statistical methods accordingly. Reader will get to know the fundamental concepts which will help them to apply statistical methods appropriately in real life situations. This book will also enable them to understand the rationale behind the methods applied. As all the examples are related to pharmaceutical sciences, this will help them to understand statistical concepts easily as well as importance of it.

An old joke goes, "there are three kinds of commonly recognised untruths: Lies, damn lies and statistics". Statistics are often used to deceive the public because most people do not understand how it works. This book will help readers to acquaint themselves with the basics of statistical analysis and protect them from being misled.

The book stems from academic material developed for workshops and pre Ph.D. course work at BV Patel Pharmaceutical Education and Research Development

(PERD) centre, Ahmedabad. I hope this book will extend learning beyond the limitations of seminar / workshop or a semester course. Of course only a limited number of topics are covered in the book, and your inputs regarding topics for further inclusion are always welcome.

I would like to particularly thank Dr Anisha Pargal who has been a constant source of inspiration, also Dr Shivprakash for his continuous encouragement and motivation. I am grateful to all the scientists / student participants of my workshops and Ph.D. course work. Their enthusiasm and interest in statistical methods was my motivation to write this book. Finally, I thank my husband, Vinay and daughter, Shuvi who have always been supportive of my ideas, practical or otherwise.

Shubha Rani

Contents

Chapter 1

Introduction

WHAT STATISTICS MEANS TO DIFFERENT PEOPLE

Different people have different opinion about statistics or statistician such as
- Statisticians spend their time in compiling tables of numbers
- Statisticians are a more glamorous version of accountant - a caretaker and manipulator of figures

The worst is "Statisticians are a type of accountant without charisma".

But thank God, there are few who think that statistics is a way of helping to make decisions, creating order out of a chaos of numbers (data), often a divinity of otherwise un-interpretable data

Actually the word "statistics" is used in two ways as follows:

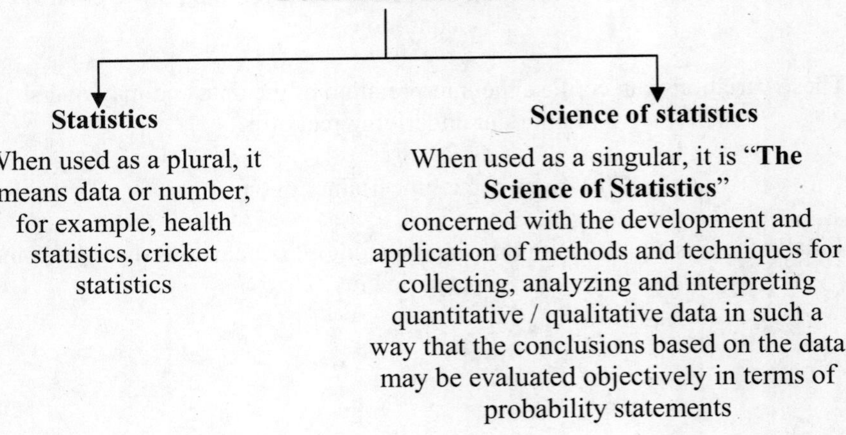

Definition of Statistics

Statistics

When used as a plural, it means data or number, for example, health statistics, cricket statistics

Science of statistics

When used as a singular, it is "**The Science of Statistics**" concerned with the development and application of methods and techniques for collecting, analyzing and interpreting quantitative / qualitative data in such a way that the conclusions based on the data may be evaluated objectively in terms of probability statements

WHY STATISTICS IS USED

Once the experiment is performed, the conclusion can be drawn directly from the experiment based on the outcome(s). The question is

- Why statistics should be used in experimental sciences, such as pharmaceutical / medical science.
- What is the need of statistics in pharmaceutical / medical sciences?

To understand the need of using statistics in pharmaceutical and medical science, let us consider few examples:

- It is very difficult to have tablets of exactly same weight in a batch. The weights will be similar but not identical. If we weigh the tablets in a batch, the weight will not be one number but may be in some range, for example, 1 – 1.05 gm.
- It is widely known through clinical trials that if the same treatment is given to the same subjects twice, the outcomes may be similar but not identical.
- If a drug is compared with placebo, there is some response for placebo also.
- Different patients may respond very differently to exactly the same dose of a drug.
- It is fairly unusual to find that a patient's / subject's pulse rate or blood pressure is exactly same when the measurement is repeated, even if only a few minutes later.
- A repeated laboratory investigation rarely gives exactly the same results. Even repeated analyses of the same sample will give a range of different results, albeit a narrow one.

Then the question is what should we conclude from this type of data? In the above mentioned first example, the weight of the tablets can be taken as 1 or 1.05 gm or any other value in between.

These examples imply that

⇓

There is some variation which can not be avoided even though we control all the conditions

⇓

These variations can confuse the interpretation of the data and may overshadow important underlying real effects

⇓

This gives rise to uncertainty in outcome

⇓

Finally there is a problem in drawing meaningful conclusions directly from the experiments

So the answer to our questions "**Is the application of the statistical methods to the subject-matter of pharmacy / medicine an important stage in the development process?**"
is
"**Yes, application of statistical methods in pharmaceutical / medical sciences is useful as we desire to draw conclusions about a group of measurements in the presence of unavoidable variability.**"

Another reason of using statistics in pharmaceutical / medical sciences is to **infer about a bigger group of units / subjects** based on the **data on a smaller group in hand**, for example,

- Only weighing few tablets from a whole batch, we extrapolate the results for the whole batch.
- Only performing dissolution test with few tablets, we infer about the dissolution profile for the whole lot of tablets.
- We may be interested in blood glucose levels in children of a certain age, we infer about this based upon the data on a small group of children of the desired age.

There are so many such examples exist in the area of pharmaceutical / medical science. Consequently, statistics is widely used in the area of pharmaceutical / medical sciences to draw meaningful conclusions.

However, always remember that statistical techniques have proven to be a great help in making wise decisions in the presence of uncertainty and also in extrapolating from a small group of units (sample) to a larger group of units (population), only if those are applied **timely** and **appropriately**.

To summarize

- Successful collaboration demands that the statistician should learn all he/she can of the problem at hand and the experimenter all should learn he/she can of the statistical approach

It is easy to lie with statistics, but it is easier to lie without it.

— Prof. Frederick Mosteller

Chapter 2

Population and Sample

Before going into more detailed statistical methods, let us have the basic understanding of population and sample.

POPULATION

A population from a statistical point of view may be loosely defined as the totality of the elements / units that have one or more properties in common.

For example,
- Population of a specific formulation produced by a manufacturer
- Population of a certain bacteria
- Group of children of a certain age is a population if we are interested in blood glucose levels in children of that age
- Population of patients with a certain disease

Population is an entire group about which some specific information is indicated i.e., population consists of data with some clearly defined characteristic(s)

Finite and Infinite Population

Population is of two types, finite or infinite:

Finite: As the name suggests when number of units in a population is finite, population is known as finite population. If one talks of the population of ages of men traveled to moon, it is a finite population. If we talk about some rare disease, population of people with that disease will be a finite population.

Infinite: When number of units in a population is infinite, population is known as infinite population. In all the above mentioned examples, population are infinite.

"Real or Existing" and "Imaginary or Hypothetical" Population

Real or existing: When population actually exists at the time of experiment, it is known as "Real or existing". In all the above examples, it is real or existing.

Imaginary or Hypothetical: A population does not exist all the time. Imagine an experiment in which a food supplement is administered to 40 guinea pigs and the sample data consists of the growth rates of these 40 pigs. Then the population about which the conclusions might be drawn is all the guinea pigs who would have been administered the same food supplement under identical conditions. Such a population is said to be "imaginary" or "hypothetical".

SAMPLE

Samples are usually a relatively small number of observations taken from a population.

Why Samples Are Required

Samples are required because
* It is possible to obtain all the measurements in a small (finite) population. But for a big (infinite) population, it is not a practical solution as it will take lots of time as well as money.
* A study of entire population is impossible in most of the situations for example, if the study process destroys or depletes the item being studied. Experiments for testing for hardness of tablets, dissolution time for a new formulation, toxicity experiments involving animal sacrifice, etc., belong to this category.
* The population may not exist i.e., it is imaginary or hypothetical.

In all such situations, the only alternative is sample study. Therefore, we take a subset of population. This subset is known as sample. From the characteristic of samples, conclusions can be drawn about the characteristic of population from which the sample is drawn.

Advantages
* Sample results are quick and less expensive at the same time more accurate.
* If samples are properly selected (i.e. samples are the true representative of the population), probability methods can be used to estimate the desired parameter and its error also, i.e., we can estimate the confidence in the calculated parameter.

- It is this aspect of sampling that permits to make probability statements about the observations / results in a study.

Disadvantages

- The variable is measured on sample, hence a substitute for the variable of population we really care about.
- Our measurements may be made or recorded incorrectly, for example, assays may not always measure exactly the right thing.
- When sample is not the true representative of the population. This disadvantages is due to sampling methods:
    - The population we really looking for is more diverse than the population from which data were sampled i.e. if sampling method is not proper. For example, if interest is in pharmacological effect of a drug in males and females both, but sample is taken only from male subjects.
    - The data is collected as a "convenience sample" rather than a random sample. For example, if it is known that a particular species mimics a particular disease model for humans, but another species is taken because of availability of that species.

Table: Some examples of population and sample

Population	Sample
Batch of ascorbic acid injection B.P.C. (1 Lac Ampoules of 2ml)	20 injections taken for leak test
Batch of tablets	20 tablets taken for estimating the content of drug
Sprague-Dawley rats	100 rats selected to test possible toxic effects of a new drug candidate
All patients with a particular disease	Selection of few patients to participate in a clinical study
Persons with diastolic blood pressure between 105 and 120 mmHg	120 Patients with diastolic blood pressure between 105 and 120 mmHg to compare two antihypertensive agents

There was the high-ranking officer in WWII who spent months counting all the bullet holes on the returning bombers, then did a big presentation on how those areas should have armor added. At the end of his presentation a lower-ranking officer asked 'Shouldn't we, instead, add more armor to those areas that are only lightly holed? After all, this sample represents only the planes that came back'.

It is clear from the above discussion that the objective of drawing samples is to estimate the characteristic of the population in which we are interested. Hence sample should be a true representative of the population to get an accurate and precise estimate of the required characteristic.

ACCURACY AND PRECISION

Accuracy: Accuracy refers to the closeness of an individual observation or mean from a sample to the true value of the parameter of the population for which the experiment was conducted or sample is taken.

Precision: It indicates how close estimates (For example, mean) are to each other calculated from various samples obtained from a population or Extent of variability of a group of measurements observed under similar experimental conditions

We try to understand these concepts by the following examples:

Example: Suppose the true value of the parameter which we want to estimate from the sample, is represented by the centre of the circle (refer the following figures) and arrows represent the various estimates of the true value of the parameter.

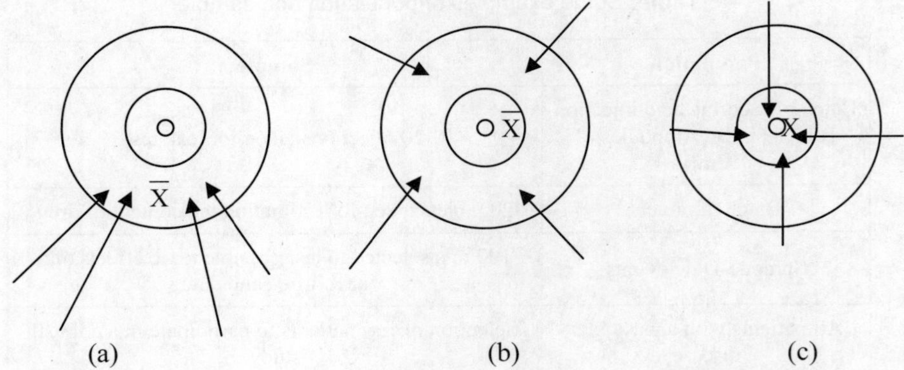

(a)   (b)   (c)

Figure: (a) represents a situation in which the estimates based on different samples are closely distributed among themselves but their average ($\overline{X}$) is not close to the true value of the parameter. This represents the case of less accurate but precise situation, (b) represents a situation in which the estimates based on samples are not closely distributed among themselves but their average is close to the true value of the parameter. This represents the case of accurate but not precise situation, (c) represents a situation in which the estimates based on samples are closely distributed among themselves and their average is also close to the true value of the parameter. This represents the case of accurate and precise situation.

Example: Suppose there is a group of numbers (94, 96, 98, 100, 102, 104 and 106) with mean 100. It can be assumed as a population with mean 100. We choose two samples of size 3 as follows: Sample 1: 96, 98 and 106; Sample 2: 98, 100 and 102.

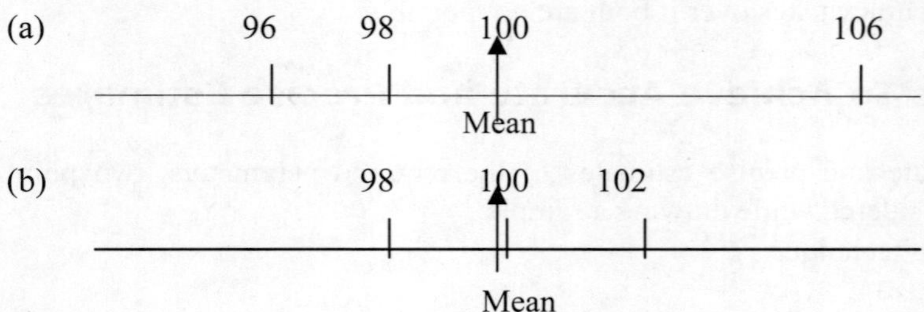

Figure: (a) represents the sample (96, 98 and 106), which has mean 100 but the observations are far apart, (b) represents the sample (98, 100 and 102), which also has mean 100 and the observations are in close proximity.

Sample of Figure (a) provides an accurate but less precise estimate of mean, whereas sample of Figure (b) provides an accurate and also precise estimate of me**an.**

Example: Following figures give the dissolution profile and plasma concentration time profile of the two formulations, I and II. Figure (A) represents the dissolution profile of both the formulations in media 1 and Figure (B) represents the dissolution profile of both the formulations in media 2. Figure (C) is the in vivo profile of the same two formulations.

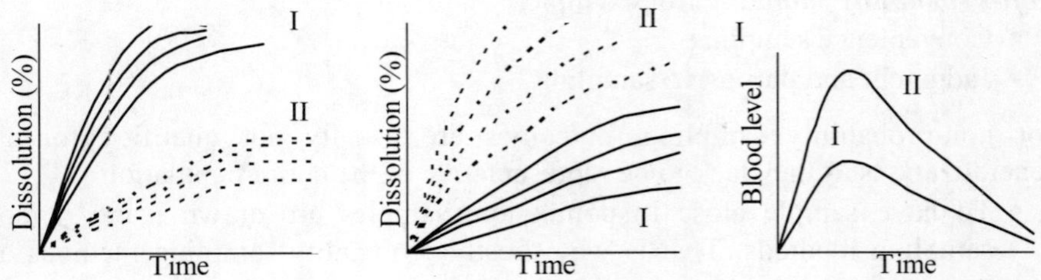

Figure: (A) represents that dissolution profile of formulation I is better than dissolution profile of formulation II and Figure (B) represents that dissolution profile of formulation II is better than dissolution profile of formulation I. Figure (C) infers that formulation II is more bioavailable compared to formulation I.

These figures imply that dissolution profiles in media 2 (Figure (B)) represents the in vivo profile (Figure (C)) correctly. This example infers that though dissolution profiles in media 1 (Figure (A)) are very precise for both the formulations I and II, dissolution profiles in media 2 is in line with the in vivo data. Therefore, even though the precision is important, in this example if we have to choose between the two media, media 2 will be preferred.

The above examples imply that both these ideas i.e. accuracy and precision are important and we aim for a precise and accurate estimate of the parameters. But in general, accuracy takes over if both are not possible.

Processes To Achieve Accurate and Precise Estimates

To get accurate and precise estimates of the required parameters, two points should be considered while drawing a sample:
- Sampling technique
- Sample size

Sampling technique
A central question of statistical conclusions is - How representative (of population) is a sample likely to be?

Appropriate sampling techniques / methods should be applied to get a true representative of the population of interest. There are many sampling methods available in the literature. The techniques of sampling may be broadly classified as follows:

Probability sampling, for example,
- Simple random sampling
- Stratified, cluster or systematic sampling
- Multi-stage and multi-phase sampling

Non-Probability sampling, for example,
- Convenience sampling
- Judgment and purposive sampling

For non-probability sampling, outcomes are usually not qualified for any generalizations as they lack to be representative of the entire population.
- To have sample close to population, samples are drawn using random sampling methods. That is why mostly probability sampling methods are used as far as possible.
- A randomly chosen sample usually will not exactly resemble the population from which it was drawn. The discrepancy between the sample and the population is called sampling error.
- The quantification of sampling error is a major contribution which statistical theory has made to scientific thinking.

Examples
Probability sampling:
- For Sterility testing, sample should be taken using simple random sampling as specified in monograph of British Pharmacopoiea Commision (B.P.C.).

Non-Probability sampling:
- To check bleeding in blister packs of marketed ethamsylate tablets, if sampling is done in a particular area of a city only. This is convenience sampling.

Many books are available on sampling techniques / methods. One can refer these books to know about sampling methods in detail. For example: Sampling Techniques by W G Cochran.

However, there may be scenario when random sampling methods are used, but the procedure used is biased or include no randomness. Few examples are as follows:

Example: A biologist plans to study the distribution of body length in a certain population of fish in a lake. The sample are collected using fishing net. Smaller fish can more easily slip through the holes in the net. Hence smaller fish will be underrepresented in the sample and the sample mean length will be an overestimate of the population mean length.

Example: A neuroanatomist plans to measure the size of individual nerve cells in cat brain tissue. In examining the tissue specimen, the investigator must decide which of the hundreds of cells in the specimen should be selected for measurement. Some of the nerve cells are incomplete because of the microtome cut through them when the tissue was sectioned. If the size measurement can be made only on complete cell, a bias may arise because the smaller cells had a greater chance of being missed by the microtome blade.

When the sampling procedure is biased, the sample parameter is a poor estimate of the population parameter.

Example: An agronomist plans to sample beet roots from a field in order to measure their sucrose content. If all the specimens are from randomly selected small area of the field, it would produce too homogeneous (less variable) a sample. Even though random sampling is done, a kind of nonrandomness is introduced.

In order to check applicability of the random sampling models, one needs to ask
- **Whether the sampling procedure is biased**
- **Whether the sampling procedure is adequately reflect the variability in the population.**

Sample size
The very first question which is asked by experimentalist is "How big a sample should be to estimate the desired parameters accurately and precisely?"

Let us see the following situation:

Suppose 100 samples (of size 5, 10, 20 and 50 each) are drawn from a population having mean = 4.50, standard deviation (sd) = 2.87. Mean is calculated for each sample separately i.e. there will be 100 sample means for each sample size. Then we classify these 100 means for each sample size as follows:

Table: Distribution of 100 means for each sample size (n)

Range of mean values	Number of means in the range (n= 5)		Number of means in the range (n= 10)		Number of means in the range (n= 20)		Number of means in the range (n= 50)	
0.75-1.25	1		-		-		-	
1.25-1.75	1		-		-		-	
1.75-2.25	4		1		-		-	
2.25-2.75	2		2		-		-	
2.75-3.25	12		5		2		1	
3.25-3.75	15		8		9		5	
3.75-4.25	12		16		24		22	
4.25-4.75	10	}39	26	}58	31	}77	45	}91
4.75-5.25	17		16		22		24	
5.25-5.75	8		15		10		3	
5.75-6.25	6		8		2		-	
6.25-6.75	7		3		-		-	
6.75-7.25	4		-		-		-	
7.25-7.75	1	-	-	-	-	-	-	
Total number of means	100		100		100		100	
Total observations	500		1000		2000		5000	
Grand mean	4.43		4.61		4.50		4.48	

Table: Summary of means of above data for each sample size

Sample size	The mean of the 100 means	The sd of the 100 means	$\dfrac{sd}{\sqrt{n}}$
5	4.43	1.36	0.61
10	4.61	0.91	0.29
20	4.50	0.61	0.14
50	4.48	0.44	0.06

The above results imply that initially increasing the sample size improves the accuracy as the mean converges towards the true population mean (4.50). But after a particular sample size, increasing the sample size does not improve the accuracy of the parameter though the standard deviation decreases, i.e. precision improves. Actually, sample size of 20 gives an accurate estimate (4.50) which is actually the population mean with fairly good precision. Further increasing the sample size, the accuracy decreases as mean became 4.48. Also number of means close to true mean of 4.50 are dependent on the sample size i.e., means of only 39 samples were in the range 3.75 to 4.25 when sample size was 5, it increased to 58 for sample size 10 and so on. It implies that chances of getting closer to the true mean is higher with larger sample size. But we have to keep the balance between the sample size and accuracy & precision depending upon the objectives.

Sample size as well as Sampling method is equally responsible for the reliable outcome of a study

To summarize

- Samples are required to study the population of interest
- Samples should be true representative of population
- Bigger does not always mean better or more powerful in making inferences
- For this reason, investigators must plan the sample size appropriately for their study prior to begin research

Birthdays are good for you. Statistics show that the people who have the most live the longest.

— **Father Larry Lorenzoni**

Chapter 3

Descriptive Statistics – Tabular and Graphical Presentation

INTRODUCTION

- Whenever we have data, the first step is to describe the data at hand in some concise way to make that easily understandable.
- Illustrating the data in a meaningful way is very important.
- But ironically, researchers spend a lot of time performing the experiments but then present them poorly which can mislead us.

The branch of statistics known as "Descriptive statistics" deals with the presentation of data using
- Tabular and graphical presentation
- Numerical measures

In this chapter, we will discuss tabular and graphical presentation of data.

TYPES OF DATA

Population is an entire group with some specific characteristic(s), i.e., it consists of data with some clearly defined qualities / attribute(s). These attributes can be of two types and hence data is of two types:
- Qualitative
- Quantitative

Qualitative Data

Qualitative data means it is represented by quality but not by number, e.g., whether a tablet is defective or not, colour of eyes, etc. However, qualitative data can be converted into quantitative data. In the above examples,

- If tablet is defective its value can be defined as 1, otherwise 0.
- If colour of eyes is black, give 1, blue give 2 and so on.

Quantitative Data

Quantitative data means it can be represented by numbers, e.g. weight of tablet, dissolution time of tablet, % dissolution of tablet, height of 5 years old children, etc.

VARIABLE

The above discussed qualities / attributes are known as "Variables". For example,

- If we are looking for weight of tablets, weight is the variable.
- If we are interested in dissolution time of the tablets, dissolution time is the variable.
- If we are concerned about the defective capsules in a batch, number of defective capsules is the variable.

Variables are of two types:
- Discrete variable
- Continuous variable

Discrete Variable

Discrete variable is one which can take only discrete values.
Examples:
- The number of defective tablets in a lot of manufactured drugs
- Patients getting cured or not after receiving a treatment
- A person is vaccinated or not
- Blood group of people
- T_{max} in pharmacokinetic / bioequivalence studies

Always remember a discrete variable can be qualitative (examples 2, 3 and 4) as well as quantitative (examples 1 and 5).

Continuous Variable

Continuous variable is one that can assume any value in a certain range.
Examples:
- Weight of tablet
- Drug content in capsule
- Blood glucose concentrations in children of a certain age
- The height or weight of students in a class
- Systolic blood pressure (SBP) and diastolic blood pressure (DBP) of adults

Important: any continuous variable can be converted into discrete variable. For example, if we talk about number of tablets having weight less than 1g instead of weight itself, it becomes discrete variable though weight is the continuous variable.

TABULAR PRESENTATION OF DATA

Tabular presentation of data is performed by
  • Frequency tables
  • Cumulative frequency tables
To understand the methods, let us consider the following example:

Example: Following are the serum cholesterol changes (from baseline) (in mg/dL) of 43 patients after administration of a drug tested for cholesterol lowering effects

17, -12, 25, -37, -29, -39, -23, -22, 0, -22, -63, 33, -31, -63, -12, -49, 15, -3, 3, -50, -7, 16, -11, -38, -17, 0, -9, -21, 1, 2, -30, -32, -34, -14, -18, 5, 16, 24, -6, -49, -8, -49, -37.

Summarize the data using frequency table and cumulative frequency tables.

Solution: Steps to prepare frequency table:
1. First we classify the data in various groups. As in this example, -64 ~ -44, -44 ~ -24, -24 ~ -4, -4 ~ 14, 14 ~ 34. These groups are known as "Classes". The first value of the class is known as "Lower limit" and the second value of the class is known as "Upper limit". Here -64 is the "Lower limit" and -44 is the "Upper limit" of class -64 ~ -44.
2. We write the number of observations falling in each class. One should clearly define if lower limit is included in the class or upper limit. In this example, 6 observations are in the range -64 ~ -44, 9 observations are in the range -44 ~ -24 and so on. This "number of observations" is known as "frequency".

Table: Frequency distribution of serum cholesterol changes

Class of serum cholesterol changes (from baseline) (in mg/dL)	Number of patients (frequency)
-64 ~ -44	6
-44 ~ -24	9
-24 ~ -4	14
-4 ~ 14	7
14 ~ 34	7

There are few rules for preparing frequency tables:
- The classes should be exhaustive, i.e., each value should be included.
- The limit for each class must agree with the accuracy of the raw data, i.e., if the values are recorded up to first decimal place, intervals should be like 1.2 – 2.3 not 1.20 – 2.30.
- Number of classes should be between 10 and 20.
- Width of **class interval should be equal.**
  Last two rules are known as thumb rules, i.e., these are recommendation but not compulsion.

Cumulative Frequency table is the sum of the frequencies of the data "up to a required level" or "above a required level". Hence it can be prepared in two ways depending on the objectives and can be prepared using frequency table.

Steps to prepare cumulative frequency table (up to a required level):
1. We take only the upper limit of the class and find out how many observations are less than this value. As in this example, how many observations are < -44, how many observations are < -24, how many observations are < -4, and so on. This is started from the first class of the frequency table.
2. Number of observations less than the upper limit can be obtained by summing the frequencies of the classes as done in the following table. In the case of cumulative frequency table, the "number of observations" is known as "Cumulative frequency".

Table: Cumulative frequency table (up to a required level) of serum cholesterol changes

Serum cholesterol changes (from baseline) (in mg/dL)	Cumulative frequency
<-44	6
<-24	6+9=15
<-4	6+9+14=29
<14	6+9+14+7=36
<34	6+9+14+7+7=43

Steps to prepare another type of cumulative frequency table (above a required level):
1. We take the lower limit of the class and find out how many observations are more than this value. As in this example, how many observations are >14, how many observations are >-4, how many observations are > -24, and so on. So this table starts from the bottom of the table if prepared using frequency table.

2. Number of observations more than the lower limit can be obtained by summing the frequencies of the classes as done in the below table.

Table: Cumulative frequency table (above a required level) of serum cholesterol changes

Serum cholesterol changes (from baseline) (in mg/dL)	Cumulative frequency
>-64	6+9+14+7+7=43
>-44	9+14+7+7=37
>-24	14+7+7=28
>-4	7+7=14
>14	7

Example: The data was collected for the granule size distribution (in mm) and presented in frequency table as follows:

Granule size range (mm)	Number of granules (frequency)
1-5	10
5-10	19
10-15	164
15-20	19
20-25	13

Find out how many granules are more than 15 mm in size.

Solution:

To find out how many granules are more than 15 mm in size, we prepare cumulative frequency table (above a required level).

Table: Cumulative frequency table for the above data

Size range (mm)	Cumulative Frequency
>1	10+19+164+19+13 = 225
>5	19+164+19+13 = 215
>10	164+19+13 = 196
>15	19+13=32
>20	13

From the above table it is clear that 32 granules are more than 15 mm in size.

GRAPHICAL PRESENTATION OF DATA

Methods for graphical presentation of qualitative data
- Pie chart / diagram
- Bar chart
- Proportional bar chart
- Percentage bar chart

To understand the above graphical methods of data presentation, let us see various scenarios for different types of graphical presentations

Example: A survey was conducted to see the prescription preference of oral hypoglycemic agents (OHA) in patients with diabetes along with hypertension by endocrinologists. Total number of endocrinologists surveyed was 15. Out of these, 9 endocrinologists prescribe metformin alone, 2 prescribe metformin + pioglitazone, and 4 prescribe glimapiride. Present the data graphically.

Solution:
This data can be presented by pie diagram as well as by bar chart as follows.

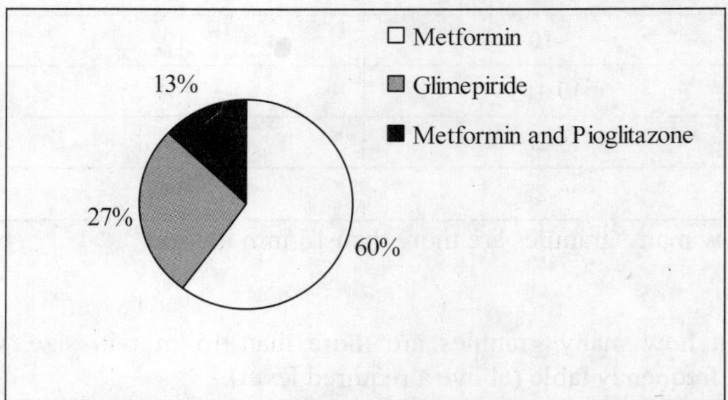

Figure: Pie diagram showing prescription preference of 15 endocrinologists for oral hypoglycemic agents (OHA) in patients with diabetes along with hypertension

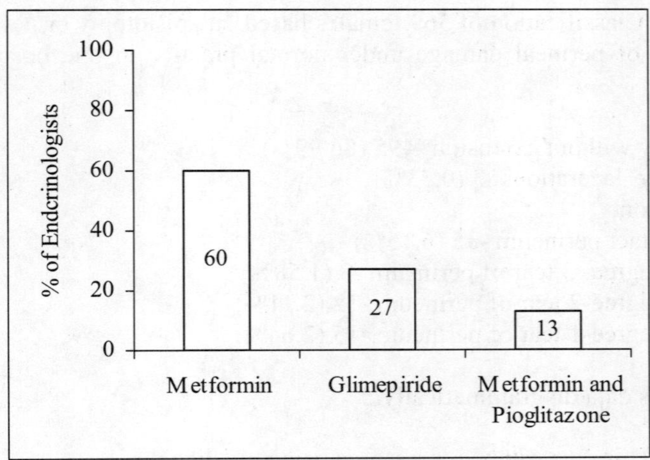

Figure: Bar chart for prescription preference of 15 endocrinologists for oral hypoglycemic agents (OHA) in patients with diabetes along with hypertension

Example: An experiment was conducted to evaluate the taste of 3 batches (Product code as Fa, Fb and Fc) of meloxicam tablets made using three different taste masking agents. Percentage of tablets having bitter and good taste are given as follows testing 6 tablets from each batch. Summarize the data graphically.

Product Code	Bitter Taste	Good Taste
Fa	67%	33%
Fb	17%	83%
Fc	17%	83%

Solution: The data can be presented using percentage bar chart.

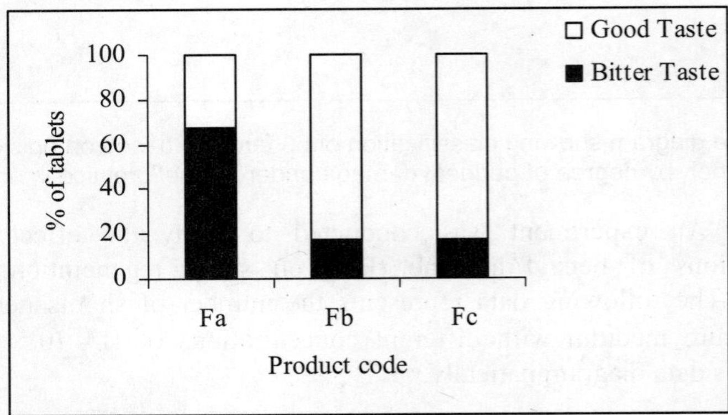

Figure: Percentage Bar chart showing % of tablets having good or bitter taste towards the meloxicam tablets obtained from three batches (n=6)

Example: Classification of 569 females based on episiotomy or not and, further, by degree of perineal damage under normal practice in one hospital is given below:

Episiotomy without extension -495 (86.99%)
Episiotomy+lacerations -3 (0.53%)
No Episiotomy
        Intact perineum -35 (6.15%)
        Degree- 3 tear of perineum -9 (1.58%)
        Degree-2 tear of perineum -12 (2.11%)
        Degree-1 tear of perineum -15 (2.64%)

Present this data diagrammatically.

Solution: This data can be represented using pie diagram.

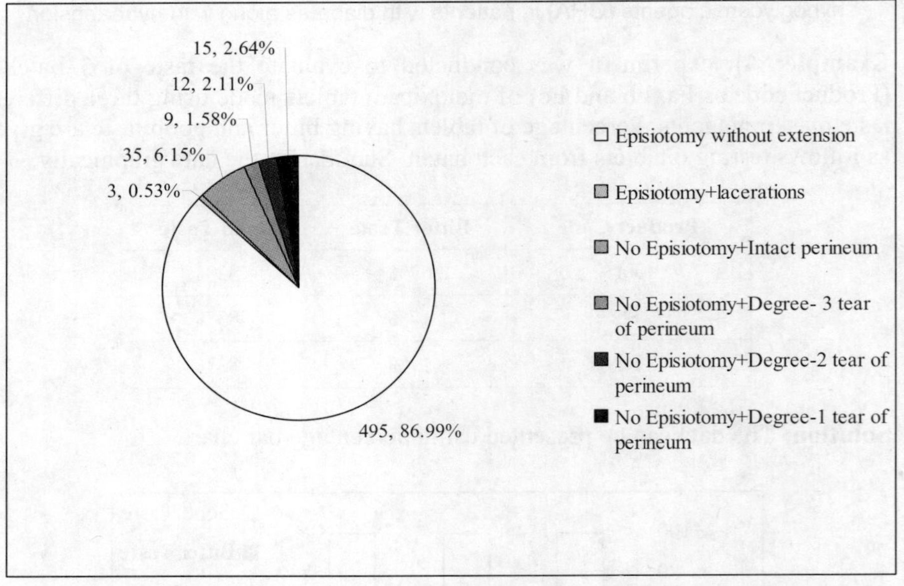

Figure: Pie diagram showing classification of 56 females based on episiotomy or not and, further, by degree of perineal damage under normal practice in one hospital

Example: An experiment was conducted to study the effect of varying concentrations of benzyl adenine (BA) on shoot regeneration of Bacopa monnieri. The following data represents the number of shoots generated in a tissue culture medium with different concentrations of BA (0.5 to 10 µM). Present this data diagrammatically.

Concentration of BA (µM)	Number of shoots regenerated
0.5	86
1	85
2	128
4	81
8	88
10	83

Solution: This data can be represented using **pie diagram** as follows:

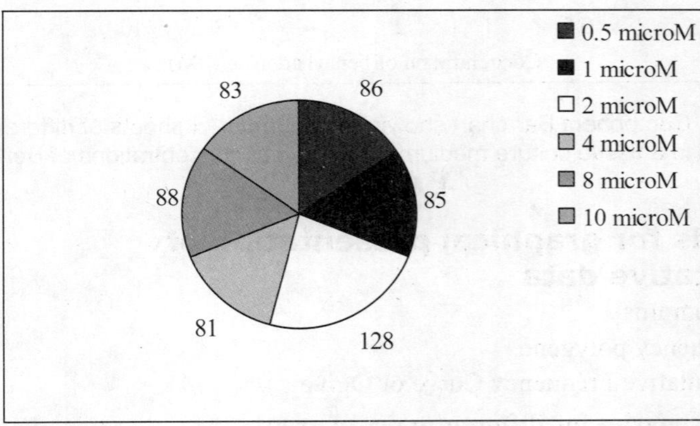

Figure: Pie diagram showing the number of shoots generated in a tissue culture medium with different concentrations of benzyl adenine (BA) (0.5 to 10 µM)

Example: The data of the above experiment was further classified as follows according to number of shoots with different lengths varying from <1 cm to 15 cm, generated using different concentrations of BA. Number of shoots generated of different lengths for various concentration of BA are given below. Present this data pictorially.

Length of the shoots	Concentration of BA (µM)					
	0.5	1	2	4	8	10
<1	5	4	19	3	6	8
1-5	67	72	96	70	75	73
6-10	11	7	11	7	7	2
11-15	3	2	2	1	0	0

Solution: This data can be represented using **component bar chart** as follows.

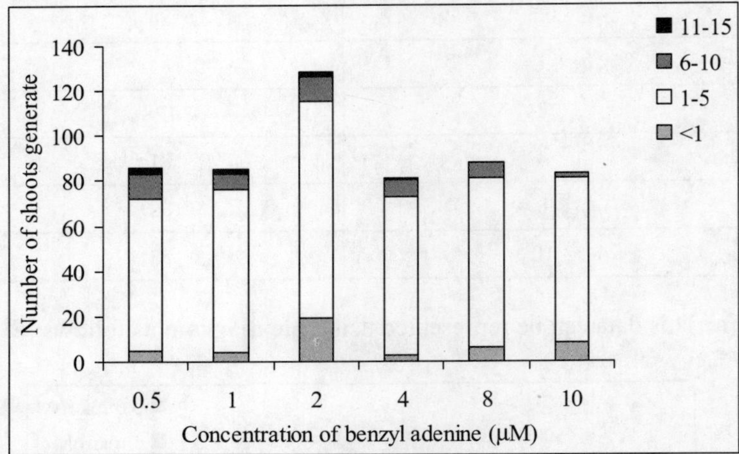

Figure: Component Bar chart showing the number of shoots of different lengths generated in a tissue culture medium with different concentrations of Benzyl Adenine (0.5 to 10 μM)

Methods for graphical presentation of quantitative data

- Histograms
- Frequency polygene
- Cumulative Frequency Curve or Ogive

Various scenarios for different types of graphs

Example: Diagrammatic presentation of the serum cholesterol change data (refer example on page no. 3.3) using **Histogram**

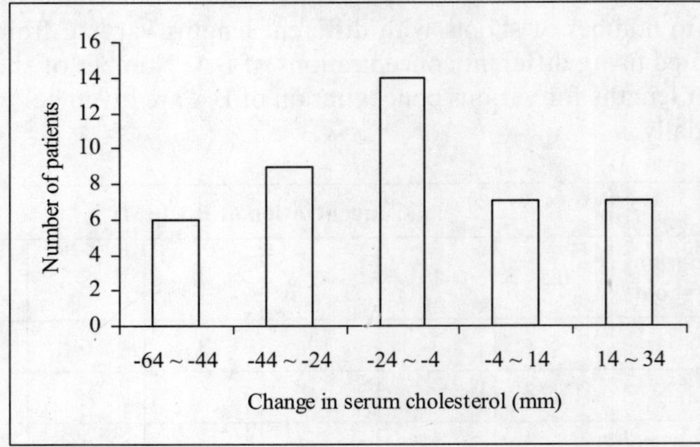

Figure: Histogram showing the serum cholesterol change in patients

Diagrammatic presentation of the serum cholesterol change data using **Frequency polygon**

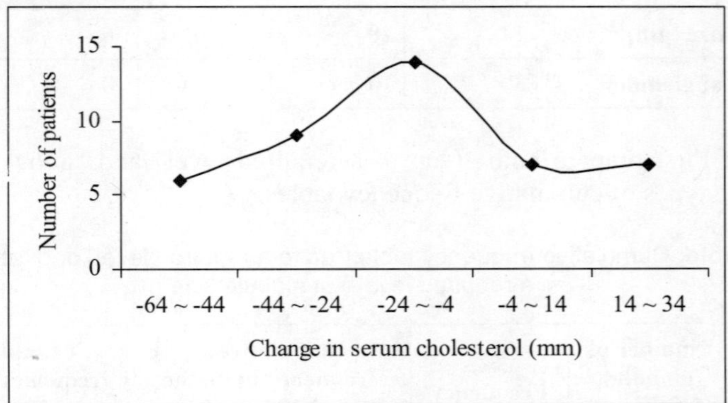

Figure: Frequency polygon showing serum cholesterol change in patients

Example: Relative dissolution of 3 batches (F(A), F(B), and F(C)) of meloxicam tablets was performed in comparison to marketed reference product and results are expressed as follows.

	Time for 100% dissolution (hr)	
	Test	**Reference**
F(A)	1.500	1.270
F(B)	1.800	1.560
F(C)	1.875	1.610

Represent the data through bar chart.
Solution: Bar chart is used to present the data graphically.

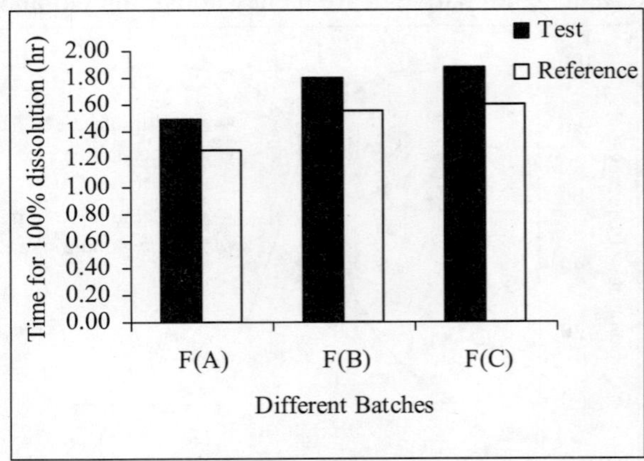

Figure: Bar chart showing relative dissolution of 3 batches (F(A), F(B), and F(C)) of test and reference tablets of meloxicam

Example: To determine mean globule diameter of an emulsion, size of 50 globules were measured. Plot the cumulative % frequency curve.

Globule size (μm)	14	28	42	56	70
Number of globules	3	10	25	10	2

Solution: First prepare both ("up to a required level" and "above a required level") the types of cumulative frequency tables.

Table: Cumulative frequency table ("up to a required level" and "above a required level") of globule size

Globule size (μm)	Number of globules (frequency)	% Frequency	Cumulative % frequency up to the globule size	Cumulative % frequency above the globule size
14	3	6	6	100
28	10	20	26	94
42	25	50	76	74
56	10	20	96	24
70	2	4	100	4

What is a cumulative frequency curve or an ogive?
A cumulative frequency curve or an ogive is an "S" shaped curve. Points on the ogive have abscissas (X axis) as the actual upper limits ("up to a required level") or lower limits ("above a required level") for curve and ordinates (Y axis) as the cumulative frequencies

Hence plot "Globule size" on x-axis and "Cumulative % frequency up to the globule size" and "Cumulative % frequency above the globule size" on y-axis as follows:

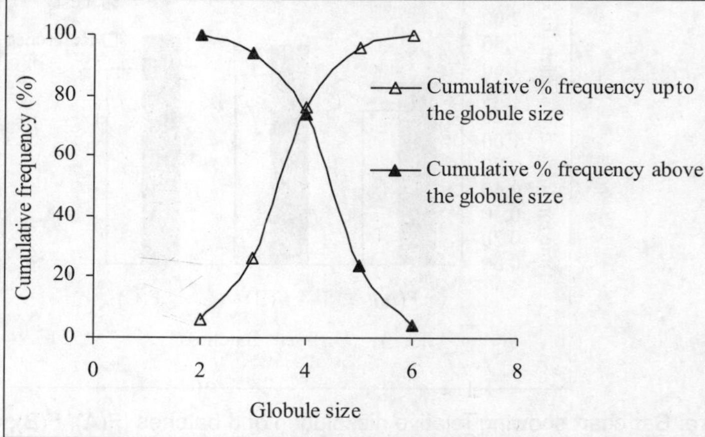

Figure: Cumulative % frequency curves of size of 50 globules

Three tricks to cheat while presenting the data graphically

Suppress the "0"

Suppose there is a weight lowering drug. So we collect data to show that it reduces the weight significantly. Collected data can be plotted in different ways. Two ways are as follows:

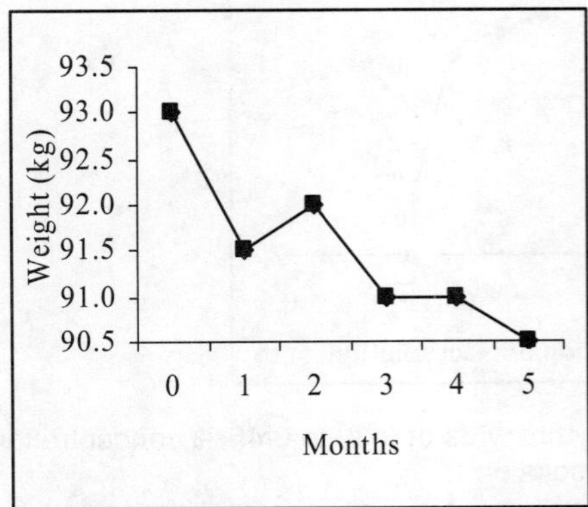

Figure: One way of plotting the months vs. weight data

Figure: Another way of plotting the months vs. weight data

At a glance, figure 1 suggests that drug is very good, whereas this is not the case with figure 2. So scale should be appropriate based on the parameter and objective.

Inflating the scale

Similarly based on our intentions, graphs can be plotted by inflating the scale. For example, blood alcohol versus accident rate can be plotted as follows:

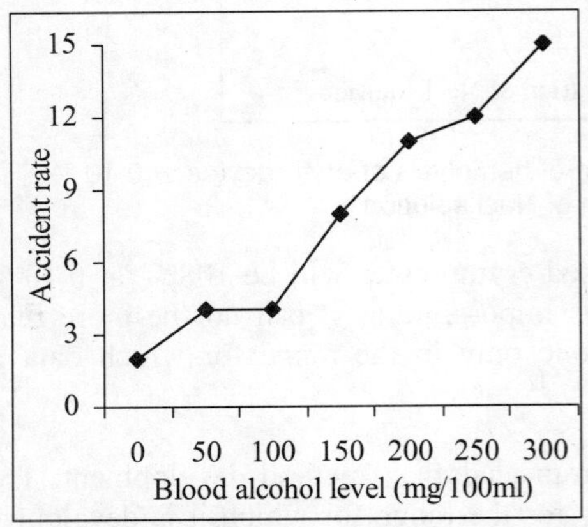

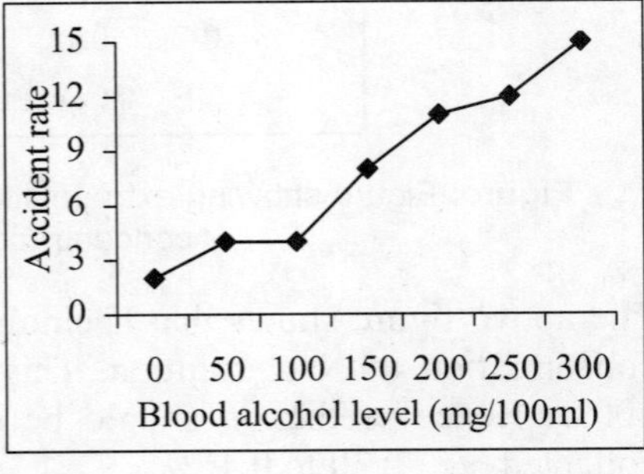

Figure: One way of plotting blood alcohol vs. accident rate

Figure: Another way of plotting blood alcohol vs. accident rate

Here again in a quick look, Figure 1 infers sharp increase in accident rates with respect to blood alcohol level compared to figure 2.

Extrapolation

Suppose data is collected for hemolysis of erythrocytes at 0.30 to 0.45 % concentration of Nacl solution. Plot of that is

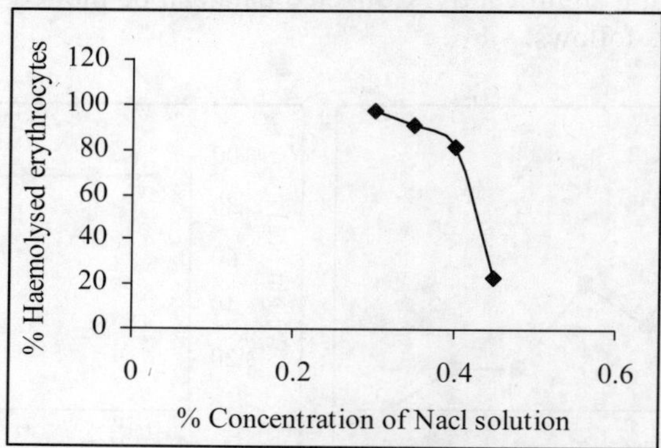

Figure: Figure showing % hemolysis of erythrocytes at 0.30 to 0.45 % concentration of NaCl solution

If we extrapolate to 0.10 % concentration of NaCl solution, plot will be as follows.

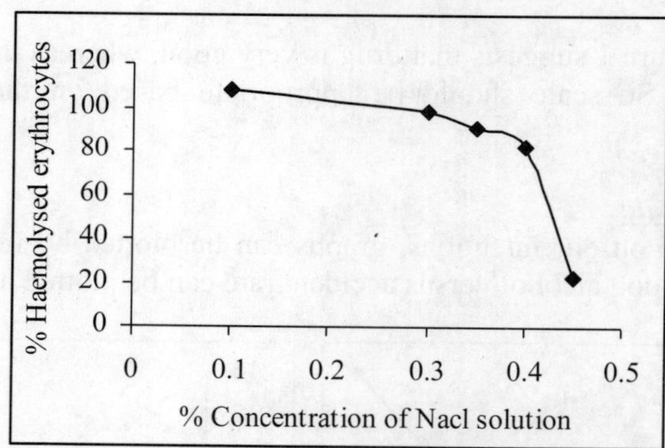

Figure: Figure showing extrapolation of hemolysis of erythrocytes at 0.10 % concentration of Nacl solution

The above figure shows that haemolysed erythrocytes will be 108% at 0.10 % concentration of Nacl solution. This is impossible as it can not be more than 100%. Hence prediction should be done only in the range for which data is collected, i.e., 0.30 to 0.45%.

A very good application of this is from analytical method development. The analytical method should be used only for the range for which it is developed. For example, if analytical method is developed for the range 10pg/ml to

1000pg/ml, it should assay only those samples which have concentrations in this range. In practical situations, the concentrations of the sample are unknown. So if concentration for sample falls out of range, dilute the sample and reanalyze. If most of the samples are falling out of range, analytical method should be refined and revalidated for the new expected range.

Three Basic Principles for Graphical Presentation

Always remember 3 principles in illustrating the results
- Heading or title: Each figure should have clearly defined heading / title
- Simplicity: Presentation of data should be simple to understand
- Honesty: If possible always give the sample size with results

To summarize:
- Tables and figures are convenient format for summarizing / presenting the data
- Greater the number of observations, greater are the economy and clarity
- Calculations also become simpler

I gather, young man, that you wish to be a Member of Parliament. The first lesson that you must learn is, when I call for statistics about the rate of infant mortality, what I want is proof that fewer babies died when I was Prime Minister than when anyone else was Prime Minister. That is a political statistic.

— **Winston Churchill**

Chapter 4

Descriptive Statistics – Numerical Measures
Measures of Central Tendency (Location) and Dispersion (Variation)

There are several numerical measures to describe a data.

MEASURES OF CENTRAL TENDENCY

It is generally seen that most of the measurements on a variable lie somewhere around the middle of the range of a data. Thus some indication of population "average" will give useful information. Such information is called measure of central tendency.

Let there is a sample where variable X takes the values $x_1, x_2, x_3, \ldots, x_{n-1}, x_n$. For example, sample of tablets with measurement on variable weight or diameter; sample of 10 year old children with measurement on variable height, etc.

Commonly used measures of central tendency in pharmaceutical sciences are:
- Mean
- Median
- Mode
- Geometric mean

Mean

Mean denoted by $\bar{x}$, is an average of all the data and is given by

Mean ($\bar{x}$)

= Sum of all the observations / Number of observations

$$= \frac{x_1 + x_2 + x_3 + \ldots\ldots + x_n}{n}$$

$$= \frac{1}{n}\sum_{i=1}^{n} x_i$$

Example: Following are the serum cholesterol changes (from baseline) (in mg/dL) of 43 patients after administration of a drug tested for cholesterol lowering effects

17, -12, 25, -37, -29, -39, -23, -22, 0, -22, -63, 33, -31, -63, -12, -49, 15, -3, 3, -50, -7, 16, -11, -38, -17, 0, -9, -21, 1, 2, -30, -32, -34, -14, -18, 5, 16, 24, -6, -49, -8, -49, -37,

Calculate average serum cholesterol change in patients.

Solution:

Mean ($\bar{x}$) $= \dfrac{x_1 + x_2 + x_3 + \ldots\ldots + x_n}{n}$

$$= \frac{17 + (-12) + (25) + \ldots\ldots + (-8) + (-49) + (-37)}{42}$$

$$= -15.7674$$

Average serum cholesterol change in patients after administration of the drug tested for cholesterol lowering effects is -15.48 units. This implies that there is an average reduction of 15.48 units in serum cholesterol after administration of the drug.

Example: A dissolution experiment of aceclofenac, 100 mg tablets, was performed using 6 tablets. The % dissolved at the end of 45 min is as follows: 85.5%, 87.1%, 88.9%, 90.1%, 86.5% and 88.4%. Find out the average % dissolved.

Solution:

Mean ($\bar{x}$) $= \dfrac{x_1 + x_2 + x_3 + \ldots\ldots + x_6}{6}$

$$= \frac{85.5 + 87.1 + 88.9 + 90.1 + 86.5 + 88.4}{6}$$

$$= 87.5$$

So the average dissolution of aceclofenac, 100 mg tablet at the end of 45 min is 87.5%.

Weighted Mean: If the values of the variable are not of equal importance, we may attach weights $w_1, w_2, w_3, ..., w_{n-1}, w_n$ to them and the weighted mean, $\overline{x}_w$ is defined as

$$\overline{x}_w = \frac{1}{\sum\limits_{i=1}^{n} w_i} \sum\limits_{i=1}^{n} w_i x_i$$

Example: Fatality rate due to a disease in various age groups is calculated. Following table gives the fatality rate and total number of patients in various age groups. Calculate the average fatality rate over all age groups.

Age in years	Fatality rate (%)	Total number of patients in the age group
<20	47.5	40
20-39	15.0	120
40-59	22.4	250
>=60	51.1	90

Solution: We calculate mean and weighted mean (total number of patients as weight) both and see which measure is appropriate for this case.

$$\text{Mean} = \overline{x} = \frac{x_1 + x_2 + x_3 + ... + x_n}{n} = \frac{47.5 + 15.0 + 22.4 + 51.1}{4} = 34.0\%$$

$$\text{Weighted mean} = \overline{x}_w = \frac{47.5 * 40 + 15.0 * 120 + 22.4 * 250 + 51.1 * 90}{(40 + 120 + 250 + 90)} = 27.8\%$$

Here the simple mean (34.0%) is quite different from the weighted mean (27.8%). Table values reveal that fatality rate is 15.0% and 22.4% for most of the patients (120+250). Hence weighted mean is comparatively closer to the true rate whereas simple mean is high due to the high fatality rate of age group of >=60 years and <20 years.
So in this case, weighted mean should be used because simple mean gives the false picture of the average fatality rate i.e. an over estimate.

Example: Following observations were made by sieve analysis of particle size distribution of pellets prepared. Calculate the average particle size.

Mean Size (µm) (x)	Number of particles (n)
213	25
300	41
470	13

Solution: Let us calculate both mean and weighted mean where number of particles is taken as weight.

$$\text{Mean} = \overline{x} = \frac{x_1 + x_2 + x_3}{n} = \frac{213 + 300 + 470}{3} = 327.67$$

$$\text{Weighted mean} = \overline{x}_w = \frac{\sum x_i n_i}{\sum n_i} = \frac{213 * 25 + 300 * 41 + 470 * 13}{(25 + 41 + 13)}$$

$$= \frac{5325 + 12360 + 6110}{(25 + 41 + 13)} = \frac{23795}{79} = 301.20$$

We can see from the above table that most of the pallets have size less than or equal to 300, so weighted mean gives the appropriate average particle size rather than simple mean.

Median

Median is the middle most value of the data.
For calculation of median,

- The given data is arranged in an increasing or decreasing order of magnitude of values.
- The middle most value in this arrangement is the median of the variable X. If number of data points is even, there will be two middle most values. In this case, median will be the average of the two middle most values.

Example: The yields of barley from 7 plots were found to be 180, 191, 175, 111, 154, 141 and 176 (in kg). Calculate the average yield.

Solution: To determine median, first arrange the data in an increasing order of magnitude as follows:

111, 141, 154, 175, 176, 180, and 191

As number of observations are 7, median is forth value which is 175 kg.

If we calculate simple mean for the above data, it is 161.14 kg. Data reveals that except 111, values are on higher side. In fact, 4 out of 7 values are equal to or greater than 175. Mean reduces because of one value 111 which is far away from remaining data. If we remove this value 111, the mean becomes 169.5 which is closer to median. Hence simple mean gives the false picture of the average yield of barley i.e. an under estimate of average yield of barley. Because of one extreme value, mean is not an appropriate measure of central tendency for this case.

Example: The distribution of particle size for bulk powder used in pharmaceutical products is often skewed. In these cases, the median is a better descriptor of the average particle size than the mean.

Example: The following data was collected for the granule size distribution (in mm):

3,4,4,3,1,3,4,4,3,2,4,3,4,2,6,4,5,4,3,4,4,3,3,3,4,1,3,4,5,4,6,4,4,3,4,5,4,3,4,2,3,4,4, 3,3,6,4,4,4,3,4,5,4,3,4,2,3,4,4,3,3,3,1,3,2,4,4,6,5,4,3,1,3,4,6,4,2,3,3,1,3,4,4,4,3,5, 3,4,2.

Find out the average granule size.

Solution:

As mentioned above, median is a better estimate for average particle size, we determine median. First arrange the data in an increasing order of magnitude. As number of observations is 89, median is forty fifth value which is 4 mm.

Example: The yields of barley from 8 plots were found to be 180, 191, 175, 111, 154, 141, 164 and 176 (in kg). Calculate the average yield.

Solution: To determine median, we first arrange the data in an increasing order of magnitude as follows:

111, 141, 154, 164, 175, 176, 180, and 191

As number of observations is 8, median will be the average of forth and fifth values which is (164+175)/2 = 169.5. So the average yield of barley is 169.5 kg.

Mode

It is the most frequently occurring value among all the observations in a sample, i.e., it is that value of the variable for which the frequency is maximum.

Example: Weight of 20 children is given below. Find out modal weight.

Weight (kg)	Number of children (Frequency)
10	3
11.5	5
13	6
14	5
14.5	1

Solution: Here the maximum number of children i.e., 6 children have weight 13 kg, hence Modal weight is 13 kg.

Example: Suppose the patients seen in a mental health clinic during a given year received one of the following diagnoses: mental retardation, organic brain syndrome, psychosis, neurosis, and personality disorder. The diagnosis occurring most frequently in the group of patients would be called the modal diagnosis.

Geometric mean

The geometric mean is given by

$$\text{Geometric mean} = (x_1 x_2 x_3 \ldots\ldots x_{n-1} x_n)^{\frac{1}{n}}$$

Geometric mean is used when data has lognormal distribution.

Example: In bioequivalence trials, pharmacokinetic parameters follow the lognormal distribution and hence geometric mean is used.

Table: A comparison of mean, median and mode

Measure	Advantage	Disadvantage
Mean	Easy to calculate Unique.	Influenced very much by very small or large values
Median	Insensitive to very large or very small values Unique.	Gives a crude estimate
Mode	Insensitive to very large or very small values.	Not necessarily unique When no value is repeated, every value is mode

MEASURES OF DISPERSION

To understand the concept and significance of measure of dispersion, let us have a look on the following example.

Frequency distribution of recorded deaths of women according to age from (1) diseases of fallopian tube, and (2) abortion, is given as follows:

Age in years as on last birthday	Death from disease of the Fallopian tube	Death from abortion
0-	1	-
5-	-	-
10-	1	-
15-	7	-
20-	12	6
25-	35	21
30-	42	22

35-	33	19
40-	24	26
45-	27	5
50-	10	-
55-	6	-
60-	5	-
65-	1	-
70-74	2	-

- Mean number of deaths due to disease of the Fallopian tube = 37.2
- Mean number of deaths due to abortion = 35.2
- So mean is almost similar for both the scenario, however, the distribution of the data is totally different in both the cases as is clear from the table.
- If only the average number of deaths of women according to age from diseases of fallopian tube and abortion is given to us, we may conclude that there is no difference between the two scenarios.

That means other than mean which is a **measure of central tendency**, we need some parameter to fully visualize the data.

The above example illustrates the importance of the spread of the data. This implies that we should know how dispersed the measurements are around the "average". Such a descriptive property of the data is called **measure of dispersion**.

Commonly used measures of dispersion are as follows:
- Range
- Variance
- Standard deviation
- % Coefficient of variation

Let variable X takes the values $x_1, x_2, x_3, \ldots\ldots, x_{n-1}, x_n$.

Range

The range is given by
Range = Maximum observation - Minimum observation
$$= x_n - x_1$$

Variance

The variance formulae are given by

$$\text{Variance (Population)} = \sigma^2 = \frac{1}{n} \sum_{i=1}^{n} (x_i - \bar{x})^2 \, , \qquad \text{for population data}$$

$$\text{Variance (sample)} = s^2 = \frac{1}{n-1} \sum_{i=1}^{n} (x_i - \bar{x})^2 \, , \qquad \text{for sample data}$$

Remember that population variance is denoted by σ^2 and sample variance is denoted by s^2.

These formulae can be simplified to

$$\text{Variance (Population)} = \frac{n \sum_{i=1}^{n} x_i^2 - (\sum_{i=1}^{n} x_i)^2}{n * n}$$

$$\text{Variance (sample)} = \frac{n \sum_{i=1}^{n} x_i^2 - (\sum_{i=1}^{n} x_i)^2}{n * (n-1)}$$

Standard Deviation

Standard deviation is square root of variance.

$$\text{Standard Deviation (Population)} = \sigma = \sqrt{\frac{1}{n} \sum_{i=1}^{n} (x_i - \bar{x})^2}$$

$$\text{Standard Deviation (sample)} = s = \sqrt{\frac{1}{n-1} \sum_{i=1}^{n} (x_i - \bar{x})^2}$$

These formulae for population and sample are same except the divisor i.e. in population case, it is n and in sample case, it is (n-1).

Standard deviation is also denoted by "sd".

The following example demonstrates why n-1 is used in calculation of sample variance or sample standard deviation instead of n.

Example: Suppose 100 samples are drawn from a population with mean = 4.50 and standard deviation = 2.87. For each sample, standard deviation is calculated using n and (n-1) in the denominator. The following table gives the frequency distribution of these sd's of all 100 samples using n and (n-1) in the denominator.

Table: Frequency distribution of 100 standard deviations (sd) using n or (n-1) in the denominator

Class interval for the sd values observed in the samples	Number of sd values in the class interval when denominator is n		Number of sd values in the class interval when denominator is (n-1)	
0.62-0.87	2		2	
0.87-1.12	2		0	
1.12-1.37	0		1	
1.37-1.62	4		1	
1.62-1.87	11	69	4	48
1.87-2.12	11		10	
2.12-2.37	13		11	
2.37-2.62	9		13	
2.62-2.87	17		6	
2.87-3.12	13		17	
3.12-3.37	8		12	
3.37-3.62	7	31	8	52
3.62-3.87	2		8	
3.87-4.21	1		5	
4.21-4.37	0		2	
Total	100	**Total**	100	
Mean of sd values	2.50		2.81	

If we look at the mean "sd" in both the cases,

- The sd with (n-1) in denominator is close to the true value 2.87. This infers that if we use n-1 as the divisor, the sample estimate is close to the population value.
- Also the sd are more evenly distributed in the case of (n-1) in the denominator (sd of 48 samples are between 0.62 to 2.87 (2.87 is the mean) and sd of 52 samples are between 2.87 to 4.37), whereas when divisor is n, sd of 69 samples are between 0.62 to 2.87 and sd of 31 samples are between 2.87 to 4.37 samples.

Keep in mind that for sample variance and sample standard deviation, the divisor is (n-1) instead of n.

Example: Following are the serum cholesterol changes (from baseline) (in mg/dL) of 43 patients after administration of a drug tested for cholesterol lowering effects

17, -12, 25, -37, -29, -39, -23, -22, 0, -22, -63, 33, -31, -63, -12, -49, 15, -3, 3, -50, -7, 16, -11, -38, -17, 0, -9, -21, 1, 2, -30, -32, -34, -14, -18, 5, 16, 24, -6, -49, -8, -49, -37,

Calculate the range, variance and standard deviation of serum cholesterol changes.

Solution:

Range $= x_n - x_1 = 33 - (-63) = 96$

Variance (sample) $= \dfrac{1}{n-1} \sum_{i=1}^{n} (x_i - \bar{x})^2 = 578.23$

Standard Deviation (sample) $= \sqrt{\dfrac{1}{n-1} \sum_{i=1}^{n} (x_i - \bar{x})^2} = 24.05$

Coefficient of Variation

Coefficient of variation is given by the following formula:

$$\%CV = \dfrac{sd}{mean} \times 100$$

Formula itself implies that
- This is nothing but the standard deviation per unit mean
- As this is free of mean value and units, this is a measure which should be used in comparing the variability of two groups having different means or units. For example, we can compare the variability of data of hardness of tablet and thickness of tablet or variability of data of blood pressure and pulse rate.

To get an idea why %CV should be used instead of sd for comparing the variability of two data sets (samples), let us see the following examples:

Example: We have two data sets, data set 1 and data set 2 as follows.
Data set 1 (in mts): 1, 2, 3, 4, and 5
Data set 2 (in cm): 100, 200, 300, 400 and 500
The objective is to see whether variability of the two data sets is same.

Solution: Let us calculate the various measure of variability using the formulae. Measures of variability are given in below table.

Table: Table showing sd, variance and %CV of data set 1 and data set 2

	Data set 1	Data set 2
	1	100
	2	200
	3	300
	4	400
	5	500
sd	1.58	158.11
Variance	2.5	25000
%CV	52.70	52.70

Data set 2 is nothing but the data as in data set 1, expressed in centimeter i.e. the multiple of 100 of all the data in the data set 1. So the variability should be same which is reflected by %CV. However, if we look at the variance, it demonstrates more variability for data set 2 compared to data set 1.

Example: An experiment was conducted to check the repeatability of an extraction procedure of a drug. A particular sample was extracted 6 times. Data obtained was as follows:

Tube No.	Area of drug	Area of internal Standard (IS)	Peak area ratio (PAR)
1	152047	833661	0.1824
2	159704	878445	0.1818
3	158117	902241	0.1752
4	133353	753334	0.1770
5	151316	831923	0.1819
6	151125	834236	0.1812

Test whether the extraction procedure is acceptable.

Solution: Any extraction procedure is acceptable if it can be repeated with minimum variability. So we calculate the mean, sd and %CV as given below.

Table: Table showing sd and %CV of data set 1 and data set 2

	Area of drug	Area of internal Standard (IS)	Peak area ratio (PAR)
Mean	150943.66	838973.33	0.1799
sd	9370.38	50980.93	0.0030
%CV	6.21	6.07	1.69
N	6	6	6

As extraction procedure will be used for various values of concentration, %CV is used to compare the variability. The %CV for the extraction procedure is 6.21% which is high, hence the extraction method is not acceptable. But if we look at the data again, we find that tube no. 4 has the value for internal standard different from the remaining values. Hence this value is removed and calculation is performed again as follows.

Table: Table showing mean, sd and %CV of data set 1 and data set 2 after removing data of tube no. 4

After deleting the tube no. 4			
	Area of drug	Area of internal Standard (IS)	Peak area ratio (PAR)
Mean	154461.88	856101	0.1805
sd	4114.08	32381.9	0.00299
%CV	2.66	3.78	1.66
N	5	5	5

Now %CV has reduced to 2.66%. As the %CV is within the permissible limits, the extraction procedure is as per the requirements. This implies that we should not reject the extraction method just based on high variability (%CV), rather we should review the data.

This is a very good example showing that statistical methods should not be used as only yardstick to draw conclusions, scientific reasoning is must.

Example: An experiment was conducted to calibrate the pipettes using methanol as test liquid. Data was obtained for 50µl and 100µl volume as follows:

	Weight (mg) for 50µl	Weight (mg) for 100µl
	0.03818	0.07745
	0.03869	0.07785
	0.03810	0.07687
	0.03826	0.07686
	0.03842	0.07668
	0.03858	0.07774
	0.03846	0.07732
Mean	0.0384	0.0773
Sd	0.0002	0.0005
%CV	0.5595	0.5947
N	7	7

The %CV is within the permissible limits, hence pipette can be used for further work.

Standard Error

- If we take various samples from the same population to estimate some parameter, for example mean, of the population, all the samples will not give the same value of the mean i.e. different samples will give different mean.
- The standard deviation of these means is known as standard error. Hence standard error is a measure of the variability of the mean.
- It is denoted by SE
- Its relation with the standard deviation (sd) of the sample is given by

$$SE = \frac{sd}{\sqrt{n}}$$

where n is the sample size.
- Smaller the standard error, more reliable the estimate is.

Formulae for various measures when data is given as frequency tables

$$\text{Mean: } \bar{x} = \frac{\sum_{i=1}^{n} f_i x_i}{\sum_{i=1}^{n} f_i}$$

Median:
- First calculate the cumulative frequency as in cumulative frequency table on page no 18.
- Divide the total frequency by 2.
- Check the class in which it lies. That is the median class.
- Median is given by

$$\text{Median} = 1 + \frac{(N/2 - C)}{f} * h$$

where l=lower limit of the median class, N is the total frequency, f is the frequency of median class, C is the cumulative frequency of the preceding median class and h is the width of the median class.

Mode:
- The class having the maximum frequency will be the modal class.
- Mode is given by

$$\text{Mode} = 1 + \frac{(f_m - f_1)}{(2f_m - f_1 - f_2)} * h$$

where l=lower limit of the modal class, f_m is the maximum frequency, f_1 is the frequency of preceding modal class, f_2 is the frequency of following modal class and h is the width of the modal class.

Variance:

$$\sigma^2 = \frac{\sum\limits_{i=1}^{n} f_i x_i^2}{\sum\limits_{i=1}^{n} f_i} - \left(\frac{\sum\limits_{i=1}^{n} f_i x_i}{\sum\limits_{i=1}^{n} f_i} \right)^2, \qquad \text{in case of population}$$

$$s^2 = \frac{1}{\sum\limits_{i=1}^{n} f_i - 1} \left[\sum\limits_{i=1}^{n} f_i x_i^2 - \frac{(\sum\limits_{i=1}^{n} f_i x_i)^2}{\sum\limits_{i=1}^{n} f_i} \right], \qquad \text{in case of sample}$$

Standard deviation:

$$\sigma = \sqrt{ \frac{\sum\limits_{i=1}^{n} f_i x_i^2}{\sum\limits_{i=1}^{n} f_i} - \left(\frac{\sum\limits_{i=1}^{n} f_i x_i}{\sum\limits_{i=1}^{n} f_i} \right)^2 }, \qquad \text{in case of population}$$

$$s = \sqrt{ \frac{1}{\sum\limits_{i=1}^{n} f_i - 1} \left[\sum\limits_{i=1}^{n} f_i x_i^2 - \frac{(\sum\limits_{i=1}^{n} f_i x_i)^2}{\sum\limits_{i=1}^{n} f_i} \right] }, \qquad \text{in case of sample}$$

Various examples of data presentation and measures of Central Tendency and Dispersion

Example: To find out the hardness of meloxicam tablets, an experiment was conducted using 6 tablets. Data is as follows:

Tablet No.	Hardness of tablets (Kg/cm^2)
1	3.5
2	3
3	3.5
4	4
5	3.5
6	4

Calculate the mean, variance and standard deviation of hardness of tablets.

Solution: Prepare the following table:

Tablet No.	Hardness of tablets (Kg/cm^2)	x_i^2
1	3.5	12.25
2	3	9
3	3.5	12.25
4	4	16
5	3.5	12.25
6	4	16
Total	21.5	77.75

$$\text{Mean} = \frac{3.5+3+3.5+4+3.5+4}{6} = \frac{21.5}{6} = 3.58$$

$$\text{Variance} = \frac{n\sum_{i=1}^{n} x_i^2 - (\sum_{i=1}^{n} x_i)^2}{n*(n-1)} = \frac{6*77.75-(21.5)^2}{6*(6-1)} = 0.1417$$

$$sd = \sqrt{\text{Variance}} = 0.3764$$

So the average hardness of tablets is 3.58 kg/cm^2, variance of the data is 0.1417 and standard deviation is 0.3764.

Example: To calculate the average number of tablets having orange-peel effect, 6 batches of aspirin tablets were tested for orange-peel effect. Samples of size 20, 10, 30, 25, 15, and 35 tablets were taken from six batches. These tablets were checked for orange-peel effect. The data is presented in the following table:

Batch No.	No of tablets Inspected In the batch (w_i)	No. of Tablets with orange peel Effect (x_i)
701	20	3
702	10	1
703	30	3
704	25	2
705	15	2
706	35	5
Total	135	16

Solution:

As sample size is different for each batch, we should calculate weighted mean rather than simple mean.

$$\text{Weighted Mean} = \overline{x}_w = \frac{1}{\sum\limits_{i=1}^{n} w_i} \sum\limits_{i=1}^{n} w_i x_i$$

$$= \frac{20 \times 3 + 10 \times 1 + 30 \times 3 + 25 \times 2 + 15 \times 2 + 35 \times 5}{135}$$

$$= 415/135 = 3.074$$

So the average number of tablets having orange-peel effect is 3.

Example: An experiment was conducted to determine the disintegration time of some dispersible tablets. Six (6) tablets were taken and results are as follows.

Tablet No.	Disintegration Time of tablets (Seconds)
1	40
2	44
3	42
4	50
5	49
6	6

Find out the average disintegration time of tablets.

Solution: As one value (6) is very low compared to other values, median should be used instead of mean.
On arranging the data in increasing order, we get
6, 40, 42, 44, 49, 50
As number of observations is even i.e. 6, median will be average of 3rd and 4th value. Hence

Median = (42+44)/2 = 43

So average disintegration time of dispersible tablet is 43 seconds.

Example: In a sample of 10 tablets, thickness of tablets was measured. Observations are as follows:

Tablet No.	1	2	3	4	5	6	7	8	9	10
Thickness (mm)	2.7	3	2.8	2.9	3	2.8	2.8	3.1	2.8	2.9

Find the modal thickness.

Solution: Prepare the frequency table as follows.

Table: Frequency table for thickness of tablets

Thickness (mm)	Frequency
2.7	1
2.8	4
2.9	2
3.0	2
3.1	1

As thickness of 2.8 mm is observed maximum time (4 times), 2.8 mm is the modal thickness.

Example: To check the uniformity of content of capsules filled on hand filling machine, a sample of 20 capsules was taken. Weight (in mg) of content of capsules is as follows:

100, 102, 106, 102, 105, 106, 108, 110, 100, 109, 106, 108, 108, 105, 102, 102, 103, 104, 101, 108

Check the uniformity of content of capsules filled on hand filling machine.

Solution: Using the formulae for mean, minimum, maximum, range, variance, sd and % CV, we get

Mean wt	104.75
Minimum Weight	100
Maximum Weight	110
Range	100-110
Variance	9.78
Standard Deviation	3.13
%CV	2.99%

As variability (%CV) is only 2.99% which is within the permissible limits, the content of capsules are filled uniformly on hand filling machine.

Example: To compare the extractable volume of injection from 2 marketed ampoules (P_{PHARMA}, Q_{PHARMA}), samples of 5 injections from the two marketed ampoules were taken and their extractable volume was measured as follows.

Extractable volume (ml) from P_{PHARMA}	Extractable volume (ml) from Q_{PHARMA}
2.1	2.5
2.1	2.2
2.2	2.2
2.1	2.3
2.1	2.4

Find out which one is a better injection based on the variability in extractable volume.

Solution: As comparison has to be done based on the variability, we will calculate %CV as follows:

	P_{PHARMA}	Q_{PHARMA}
Mean	2.12	2.32
Standard Deviation	0.0447	0.1304
Variance	0.002	0.017
% CV	2.1095	5.6200

Higher value of %CV in product of Q_{Pharma} indicates more variation in extractable volume. Hence injection from P_{PHARMA} is a better injection.

Example: At the product development stage of 100mg aceclofenac tablet, two different formulas were developed. One batch was prepared for each formula. Dissolution profiling was done for both the batches of aceclofenac tablet using a sample of 6 tablets at 3 time points. The dissolution data for both batches at 3 time points, 15, 30 and 45 min, is given below. Finalize the formula based on this data.

Time (min)	% Dissolution of Formulation A	% Dissolution of Formulation B
15	50,65,70,55,65,68	64,64,65,65,66,67
30	70,79,74,74,72,75	75,78,78,76,77,76
45	85,89,89,98,95,95	94,95,95,94,93,96

Solution: Calculate mean, variance, standard deviation, and %CV of dissolution data:

Table: Table giving the mean % dissolution, its variance, standard deviation and %CV for formulation A and formulation B

Time (min)	% Dissolution	Mean % dissolution	Standard Deviation	Variance	%CV
Formulation A					
15	50,65,70,55,65,68	62.17	7.885	62.197	12.653
30	70,79,74,74,72,75	74.00	3.033	9.200	4.099
45	85,89,89,98,95,95	91.83	4.916	24.167	5.353
Formulation B					
15	64,64,65,65,66,67	65.17	1.169	1.367	1.794
30	75,78,78,76,77,76	76.67	1.211	1.467	1.579
45	94,95,95,94,93,96	94.5	1.048	1.100	1.109

From the above table, it can be concluded that although average % dissolution for both the batches at 15, 30, and 45 min is almost similar, formulation B is better in comparison of formulation A as it has less %CV at each time point.

Here it should be noted that even variance and standard deviation have the similar pattern as %CV. This example shows that if units or mean of the samples to be compared are same, variance and standard deviation give the same result as %CV.

Example: To compare two different methods of glucose estimation (enzymatic method and chemical method), glucose estimation was done for the same sample 9 times using both the methods. Results are as follows.

Estimated value of Glucose (mg/dL) in sample using Enzymatic method	Estimated value of Glucose (mg/dL) in sample using Chemical method
99	100
91	95
94	97
97	99
93	95
95	98
96	99
97	99
99	100

Find out which one is a better method for estimation of glucose.

Solution: The method which gives less variability will be better. Hence we will calculate %CV as follows:

	Enzymatic method	Chemical method
Mean	95.67	98.00
Standard Deviation	2.69	1.94
% CV	2.81	1.98

Variability (%CV) of Chemical method is less compared to enzymatic method, so it is concluded that Chemical method is a better method.

Example: Following table shows the weights of double distilled water aspirated 10 times by a 200 µl micropipette for a fixed volume. Find out for which volume the observations are the most precise among all.

No.	Readings for 50µl Weight (gm)	Readings for 100 µl Weight (gm)	Readings for 150 µl Weight (gm)
1	0.048	0.100	0.146
2	0.047	0.099	0.147

3	0.048	0.099	0.147
4	0.046	0.101	0.147
5	0.048	0.099	0.148
6	0.047	0.102	0.148
7	0.048	0.099	0.147
8	0.047	0.099	0.148
9	0.049	0.100	0.145
10	0.048	0.100	0.147

Solution: Mean, standard deviation (SD), standard error of mean (SEM) as well as % coefficient of variation of the data are as follows:

Mean	0.0476	0.0998	0.147
SD	0.0008	0.001	0.0009
SEM	0.0003	0.0003	0.0003
%CV	1.772	1.035	0.641

As %CV is least for 150 µl, observations are more precise for 150 µl.

Example: Two batches of size 100 tablets and 500 tablets were prepared. To test the effect of increasing batch size on weight variation of tablets, samples of 20 tablets were taken from each batch. Weights of 20 tablets are as follows:

Tablet number	Weight of tablets from Batch of size 100 tablets (mg)	Weight of tablets from Batch of size 500 tablets (mg)
1	224.2	229.2
2	225	225.8
3	224.5	227.5
4	221.5	220
5	222.8	224.5
6	226.5	228.5
7	223.7	221.8
8	228.4	224.3
9	224.4	225.2

10	225.6	228.5
11	228.4	222.6
12	220.8	227.4
13	228.5	227.9
14	225	227.4
15	223.6	223
16	224.3	225.5
17	224	227.1
18	227.5	226.6
19	228	227.8
20	227.5	229.4

Find out for which size batch weight variation is less.

Solution: As comparison has to be done based on the variability, we will calculate %CV as follows:

	Batch of size 100 tablets	Batch of size 500 tablets
Mean	225.21	226
Standard Deviation	2.29092	2.607277
% CV	1.02	1.15

As % CV for both the batches are almost same, increasing the batch size does not increase the weight variation of tablets. Hence size of the batch can be increased if required.

Example: Consider the vincamine production in multiple shoot culture derived from hairy roots of vinca minor. Following are the vincamine content (units) in 20 regenerated shoots.

0.53	0.50	0.54	0.48	0.57	0.55	0.51	0.52	0.49	0.50
0.50	0.54	0.56	0.51	0.53	0.53	0.55	0.52	0.51	0.52

Summarize the data by various methods and also present graphically.

Solution:

Data presentation using frequency table: As data values vary from 0.47 to 0.58, we classify the data in 6 classes of width 0.01.

Frequency distribution of vincamine content in regenerated shoots is as follows.

Table: Frequency distribution of vincamine content in regenerated shoots with other calculations required to calculate various measures

Sr.No.	Class Internal	Frequency (f_i)	Mid point (x_i)	x_i^2	$f_i x_i$	$f_i x_i^2$
	0.47 -0.48	1	0.475	0.2256	0.474	0.2256
	0.49 -0.50	4	0.495	0.2450	1.98	0.98
	0.51 - 0.52	6	0.515	0.2652	3.09	1.5912
	0.53 - 0.54	5	0.535	0.2862	2.675	1.431
	0.55 - 0.56	3	0.555	0.308	1.665	0.924
	0.57 - 0.58	1	0.575	0.3306	0.575	0.3306
Total		20			10.46	5.4824

Mean and standard deviation of the data

$$\bar{x} = \frac{\sum_{i=1}^{n} f_i x_i}{\sum_{i=1}^{n} f_i} = \frac{10.46}{20} = 0.523$$

$$s = \sqrt{\frac{1}{\sum_{i=1}^{n} f_i - 1}\left[\sum_{i=1}^{n} f_i x_i^2 - \frac{\left(\sum_{i=1}^{n} f_i x_i\right)^2}{\sum_{i=1}^{n} f_i}\right]} = 0.0249$$

The average vincamine content in regenerated shoots is 0.523 units with standard deviation 0.0249.

Graphical representation of the the data is done using frequency curve.

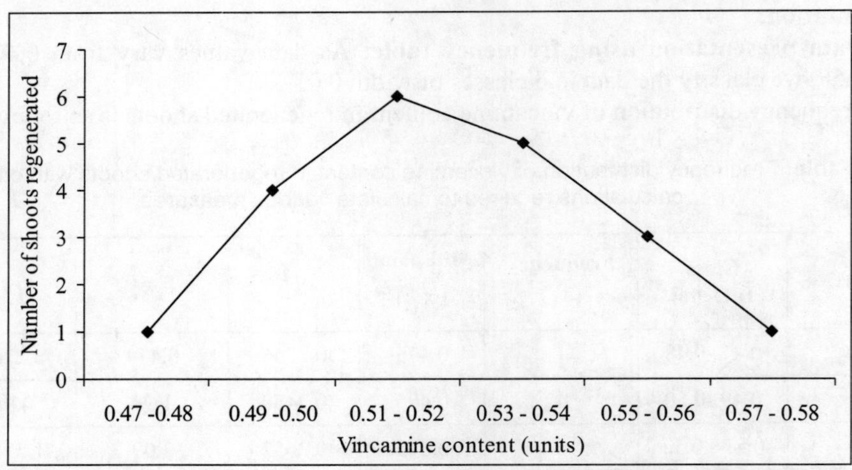

Figure: Frequency curve of vincamine content in regenerated shoots

Example: Statistical diameter of 118 particles was obtained by using microscopic method. Data is presented in the form of frequency table as follows. Summarize the data.

Size range of diameter (μm)	0.5-1.0	1.0-1.5	1.5-2.0	2.0-2.5	2.5-3.0	3.0-3.5	3.5-4.0	Total
Number of particles in each size range	2	10	22	54	17	8	5	118

Solution: Required calculations are done in the following table.

Table: Required calculations for mean and standard deviation

Size range of diameter (μm)	Mid point of size range of diameter (x_j) (μm)	Number of particles in each size range (f_j)	$f_j x_j$	$f_j x_j^2$
0.5-1.0	0.75	2	1.5	1.13
1.0-1.5	1.25	10	12.5	15.63
1.5-2.0	1.75	22	38.5	67.38
2.0-2.5	2.25	54	121.50	273.38
2.5-3.0	2.75	17	46.75	128.56
3.0-3.5	3.25	8	26.00	84.5
3.5-4.0	3.75	5	18.75	70.31
Total		118	265.5	640.89

As number of data points is large, we can use the formula for population variance.

$$\overline{X} = \frac{\sum f_j x_j}{\sum f_j} = \frac{265.50}{118} = 2.25$$

$$\sigma = \sqrt{\frac{\sum f_j x_j^2}{\sum f_j} - \left(\frac{\sum f_j x_j}{\sum f_j}\right)^2} = \sqrt{\frac{640.89}{118} - (2.25)^2} = 0.30726$$

The average statistical diameter of particles is 2.25 with standard deviation 0.3073.

Example: An experiment was conducted to determine the volume of normal saline bottles. Volume of 100 bottles was measured. Following table gives the frequency distribution of the volume of 100 normal saline bottles.

Volume range of bottles (ml)	Number of bottles
527.5-532.5	05
532.5-537.5	18
537.5-542.5	42
542.5-547.5	27
547.5-552.5	08

Present the data graphically and calculate mean and standard deviation.

Solution:

Data is presented graphically as follows:

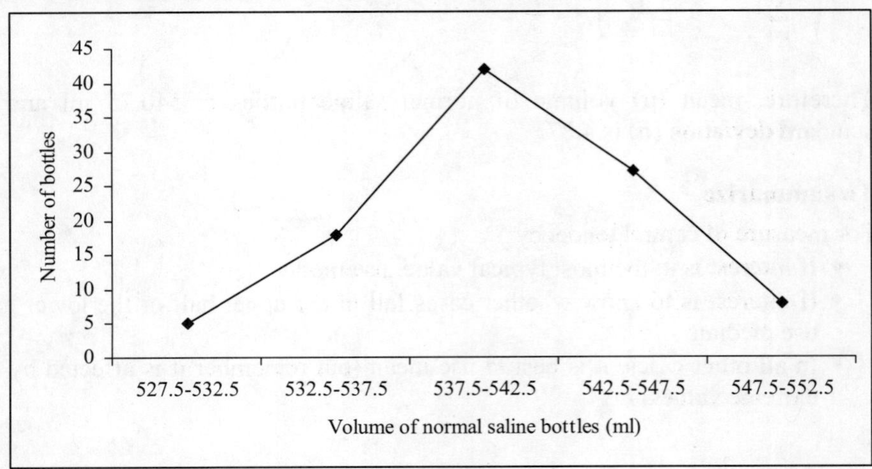

Figure: Frequency polygon showing distribution of volume of normal saline bottles

To calculate the mean and standard deviation, we perform the following calculation.

Table: Table showing the calculations for mean and standard deviation

Volume range of bottles (ml)	Mid value of the volume range (ml) (x_i)	Frequency (f_i)	$f_i x_i$	$f_i x_i^2$
527.5-532.5	530	05	2650	1404500
532.5-537.5	535	18	9630	5152050
537.5-542.5	540	42	22680	12247200
542.5-547.5	545	27	14715	8019675
547.5-552.5	550	08	4400	2420000
Total		100	54075	29243425

As number of data points is large, we can use formula of variance for population.

$$\text{Mean} = \bar{x} = \frac{\sum_{i=1}^{n} f_i x_i}{\sum_{i=1}^{n} f_i} = \frac{54075}{100} = 540.75$$

$$\sigma = \sqrt{\frac{\sum_{i=1}^{n} f_i x_i^2}{\sum_{i=1}^{n} f_i} - \left(\frac{\sum_{i=1}^{n} f_i x_i}{\sum_{i=1}^{n} f_i}\right)^2} = \sqrt{\frac{29243425}{100} - \left[\frac{54075}{100}\right]^2} = 4.87$$

Therefore, mean (μ) volume of normal saline bottles is 540.75 ml and its standard deviation (σ) is 4.87.

To summarize

For measure of central tendency
- If interest is in the most typical value, use mode
- If interest is to know whether cases fall in the upper half or the lower half, use median
- In all other cases, it is best to use mean (but remember it is affected by the extreme values)

For measure of variation
- If interest is in measuring the variability of mean, use standard error
- if interest is in comparing the variability of two groups, use %CV
- in all other cases, any measure can be used

Statistics: The only science that enables different experts using the same figures to draw different conclusions.

— **Evan Esar**

Chapter 5

Descriptive Statistics – Numerical Measures
Ratio, Proportion and Percentage

There are some other numerical measures which are used to describe data. Though not limited, the measures

- Ratio
- Proportion
- Percentage (%)

are mainly used to express qualitative results.

To understand the concept, let us consider an example of red cell volume, white cell volume and total blood volume. Our interest is to know what is the ratio of red cell volume and white cell volume, in what proportion red cell volume are in total blood volume and what is the percentage of red cell volume in total blood volume. The measures ratio, proportion and percentage are given as follows:

$$\text{Ratio of red cell volume and white cell volume} = \frac{\text{The red cell volume}}{\text{The white cell volume}}$$

$$\text{Proportion of red cell volume in total blood volume} = \frac{\text{The red cell volume}}{\text{The total blood volume}}$$

$$\% \text{ of red cell volume in total blood volume} = \frac{\text{The red cell volume}}{\text{The total blood volume}} \times 100$$

Whenever data is reported as proportions or percentages, always report the number of observations used for calculation of that percentage. Why? Let us see the following example.

Example: Suppose a surgeon claims that he/she gets success 50% of the times in the operations. If he has performed only 2 operations, this 50% gives the false picture as this is a very small sample size to draw any conclusion.

Examples of ratio, proportions and percentages:

Example of <u>ratio</u>: For the analysis of Rifampicin and Isoniazid in pharmaceutical dosage forms, a reverse phase liquid chromatography method is used. The aqueous mobile phase composition is Methanol and 0.02 M Disodium Hydrogen Orthophosphate in 3:1 **ratio,** with pH 4.5 adjusted with orthophosphoric acid.

Example: Formula for composition of a tablet is as follows:

Ingredients	Quantity per tablet (mg)
Meloxicam	7.5
alpha CD	20.75
Aerosil	1.5
MCC	170.25
Total	200

Find out the ratio of Meloxicam and alpha CD, proportion of Meloxicam and percentage of Meloxicam in the tablet. Also present the data diagrammatically.

Solution:

Ratio
= Quantity of Meloxicam / Quantity of Alpha CD
= 7.5/20.75 = 0.3614

Proportion of Meloxicam in the tablet
= Quantity of Meloxicam / Total quantity
= 7.5/200 = 0.0375

Percentage of Meloxicam in the tablet
= Quantity of Meloxicam * 100 / Total quantity
= 7.5*100/200
= 3.75%

Graphical presentation of tablet composition using Pie chart is as follows.

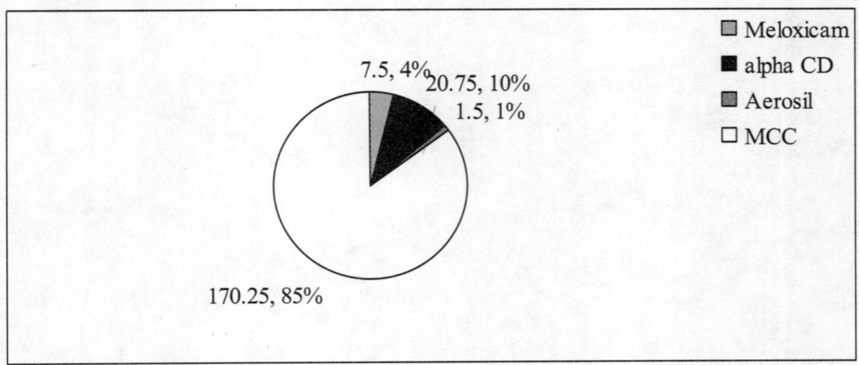

Figure: Pie chart showing tablet composition of tablet (values represent the actual quantity of ingredient and percentage of that ingredient in the tablet)

Example of proportions and percentages: Aceclofenac tablet formulation has the following formula:

S.No.	Ingredients	Qty/ batch (mg)	Composition	%Composition
		Intra granular		
1	Aceclofenac	100.00	0.5	50.00
2	Microcrystalline cellulose	83.00	0.415	41.50
3	Ac di sol	3.00	0.015	1.50
4	PVP K-30	10.00	0.05	5.00
		Extra granular		
5	Ac di sol	1.00	0.005	0.50
6	Magnesium stearate	2.00	0.01	1.00
7	Aerosil	1.00	0.005	0.50
Total		200.00		100.00

Here "Composition" is **proportion** and "%Composition" is **percentage**.

To summarize

- Unless the actual number involved is quoted, mistrust proportions and percentages as those are misused

There are two kinds of statistics, the kind you look up and the kind you make up.

— **Rex Todhunter Stout**

Chapter 6

Theoretical Distributions of Variables

WHAT IS THEORETICAL DISTRIBUTION OF A VARIABLE

- After arranging the data in a frequency distribution form, when we present the data graphically, it has some shape, refer the examples of chapters 3 and 4.
- If the frequencies are divided by total frequency, we get relative frequency distribution. An example is given in the following table.

Table: Table showing frequencies and relative frequencies

Class	Frequency	Relative frequency
-64 ~ -44	6	6/43 = 0.139535
-44 ~ -24	9	9/43 = 0.209302
-24 ~ -4	14	14/43 = 0.325581
-4 ~ 14	7	7/43 = 0.162791
14 ~ 34	7	7/43 = 0.162791
Total	43	1

Relative frequencies refer that if we perform an experiment, what are the chances of getting a particular outcome in one trial.

A theoretical distribution can be conceived as a limiting form of the relative frequency distribution of sample when sample size increases to infinity. The sample relative frequencies are known as probabilities in case of theoretical distributions.

It is clear from the above table that
- Relative frequency can never be less than 0 or greater than 1 as those are obtained by dividing the individual frequency by total of frequencies.
- Sum of all the relative frequencies is always 1 as

> 6/43+9/43+14/43+7/43+7/43
> = (6+9+14+7+7)/43
> = sum of all frequencies/ sum of all frequencies
> = 1

WHY ARE THEORETICAL DISTRIBUTIONS REQUIRED

Theoretical distributions are required
- As most of the statistical analyses proceed after the form of an appropriate theoretical distribution is decided upon. For example, if we want to compare two groups using t-test (we will discuss later in Chapter 8), data should have normal distribution.
- If we know the theoretical distribution, we can get approximate probabilities of different outcomes without performing the experiment.

HOW THEORETICAL DISTRIBUTIONS ARE DECIDED

We can usually conceive and assume the functional form of the theoretical distribution involved on the basis of experience, theoretical considerations, or information in the sample (based on relative frequency distribution). Experience with sample distributions of various measurements in biological and pharmaceutical sciences shows the trend to possess the normal distribution in most of the cases, at least approximately.

SOME COMMONLY USED THEORETICAL DISTRIBUTIONS

We will look only at the following basic distributions very commonly used in pharmaceutical / medical sciences.

Discrete distribution: These distributions are applicable when variable is discrete.
- Binomial
- Poisson

Continuous distribution: These distributions are applicable when variable is continuous.
- Normal
- Log-normal

Discrete Distribution

In discrete distributions, the distribution may look like as follows or any other form with no continuity because x assumes only discrete values:

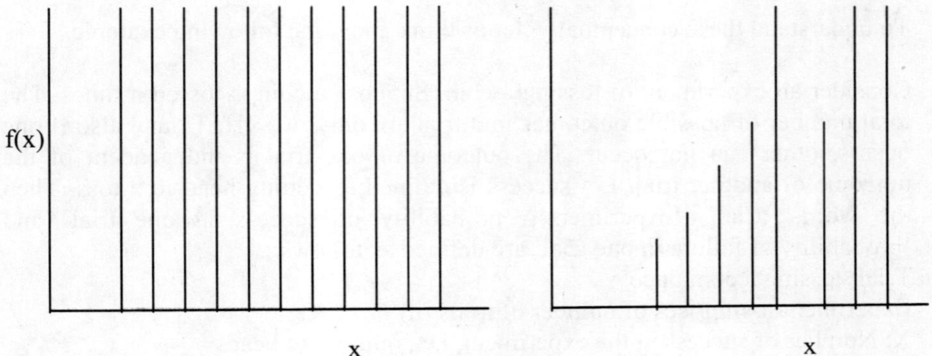

Figure: Some discrete distributions

Binomial distribution

Let X be a variable defined as the number of success in an experiment which consists of n trials. Binomial distribution can be assumed for this random variable X, if the following conditions are met in any experiment:

- There are only two outcomes of a trial which are mutually exclusive i.e. if one outcome takes place, other is not possible.
- Outcomes of one trial are independent of another trial.

Probabilities for this variable are given by:

$$P(X = x) = {}^nC_x\, p^x q^{n-x}$$

where
n = number of trials which makes an experiment
x = number of success in an experiment
p = probability of success in one trial
q = probability of failure in one trial = 1 - probability of success in one trial = 1-p

$$ {}^nC_x = \frac{n!}{x!(n-x)!} $$

This is known as Binomial probability function.

As mentioned earlier,
- Probability can never be less than 0 as those are obtained by dividing the individual frequency by total of frequencies, hence

$$^{n}C_{x}p^{x}q^{n-x} \geq 0 \qquad \text{for all x}$$

- Sum of all the probabilities is always 1, hence

$$\sum_{x=0}^{n} {}^{n}C_{x}p^{x}q^{n-x} = 1$$

To understand these conceptually, let us think about the following example.

Consider an experiment of tossing a coin. Suppose a coin is tossed n times. The total number of possible outcomes in a trial are only two {H, T} and also if one occurs, other can not occur. The outcome of one trial is independent of the outcome of another trial. Let success is defined as getting head in a toss. Then the words 'trial', 'experiment', 'probability of success in one trial' and 'probability of failure in one trial' are defined as follows:

Trial: tossing a coin once

Experiment: comprises of number of trials (n)

X: Number of success in the experiment, i.e., number of heads

Probability of success in one trial = Probability of getting head in one trial

$$= p = \frac{\text{Number of heads in a trial}}{\text{Total possible outcomes in the trial}} = \frac{1}{2}$$

Probability of failure in one trial $= q = 1\text{-}p = 1 - 1/2 = 1/2$

Applying Binomial distribution, we get

$$P(X = x) = {}^{n}C_{x}p^{x}q^{n-x} = {}^{n}C_{x}\left(\frac{1}{2}\right)^{x}\left(\frac{1}{2}\right)^{n-x}$$

If we toss only two times, i.e. n=2, possible number of heads are 0, 1 or 2, i.e., X = 0, 1, 2. Then probability of getting 0, 1 or 2 heads will be as follows:

P (Getting zero head in the experiment i.e. X =0) $= {}^{n}C_{x}p^{x}q^{n-x}$

$$= {}^{2}C_{0}\left(\frac{1}{2}\right)^{0}\left(\frac{1}{2}\right)^{2-0}$$

$$= \frac{2!}{2!0!}\left(\frac{1}{2}\right)^{2}$$

$$= 1/4$$

P (Getting one head in the experiment i.e. X =1) $= {}^{n}C_{x}p^{x}q^{n-x}$

$$= {}^{2}C_{1}\left(\frac{1}{2}\right)^{1}\frac{1}{2}^{2-1}$$

$$= \frac{2!}{1!1!}\left(\frac{1}{2}\right)^2$$
$$= 2/4 = 1/2$$

P (Getting two heads in the experiment i.e. X =2) = $^nC_x p^x q^{n-x}$

$$= {}^2C_2\left(\frac{1}{2}\right)^2\left(\frac{1}{2}\right)^{2-2}$$

$$= \frac{2!}{2!0!}\left(\frac{1}{2}\right)^2$$

$$= 1/4$$

Alternative method for solution

If instead of applying the Binomial distribution function, we calculate the probabilities directly as follows:

If a coin is tossed two times, total number of possible outcome will be head (H) in both toss, first head (H) then tail (T), first tail (T) and then head (H) or tail (T) both the times, i.e., {HH, HT, TH, TT}.

P (Getting zero head) = No of cases with zero head/total possible outcomes
$$= 1/4$$

P (Getting one head) = No of cases with one head/total possible outcomes
$$= 2/4$$
$$= 1/2$$

P (Getting two heads) = No of cases with two heads/total possible outcomes
$$= 1/4$$

We observe that the results can be obtained directly without assuming any form of distribution. But it becomes very convenient to calculate probabilities once we assume the theoretical distribution as we do not have to bother about total possible outcomes.

Example: Probability that a person will die within a month after a certain cancer operation is 18%. What are the probabilities that in three such operations, one, two or all three persons will survive?

Solution: In this example,
- There are only two possible outcomes: either person will survive or person will die. Also these two outcomes can not occur together.
- Result of one operation is independent of another operation.

As both the assumptions of Binomial distributions are satisfied, we can assume Binomial distribution and calculate the desired probabilities as follows:

Trial: one operation

Experiment: Consists of three operations, hence n =3

X = Number of success in the experiment i.e. Number of persons who survives after the experiment (3 operations), hence it can be 0, 1, 2, or 3

p = probability of success in one trial = P (survival in one trial) = 1- P (death in one trial) =1 - 0.18 = 0.82

q = P (death in one trial) = 0.18

$$P \text{ (One person will survive in the experiment i.e. } X{=}1) = {}^{n}C_{x}p^{x}q^{n-x}$$
$$= {}^{3}C_{1}p^{1}q^{3-1}$$
$$= 0.080$$

$$P \text{ (Two persons will survive in the experiment i.e. } X{=}2) = {}^{3}C_{2}p^{2}q^{3-2}$$
$$= 0.363$$

$$P \text{ (Three persons will survive in the experiment i.e. } X{=}3) = {}^{3}C_{3}p^{3}q^{3-3}$$
$$= 0.551$$

Thus the chances that only one person will survive after the experiment consisting of 3 operations is 8%, two survive is 36% and all the three will survive is 55%.

Example: Assume that a drug causes a serious side effect at a rate of three patients per one hundred. What is the probability that the side effects occurs in a random sample of ten patients taking the drug?

Solution: If a person is taking the drug,

- There are only two possible outcomes: either patient will have the side effect or patient will not have the side effect. Also these outcomes can not happen together.
- Suffering with side effects of one patient is independent of suffering with side effect of another patient.

As both the assumptions are satisfied in this situation, we can assume Binomial distribution and calculate the desired probabilities as follows:

Trial: 1 patient taking the drug

Experiment: Consists of 10 patients taking the drug, hence n =10

X = Number of persons having side effects after taking drug. It can be 0, 1, 2, 3, 4, 5, 6, 7, 8, 9 or 10

p = Probability of having side effects in one trial = 3/100 = 0.03

q = P (Not having side effects in one trial)

  =1- P (Having side effects in one trial) = 1 – 0.03 = 0.97

We wish to calculate the probability that the side effects occur in a random sample of ten patients taking the drug. This means our interest is in calculating the probability of having side effect either by 1 or by 2 or by 3 or by 4 or by 5 or by 6 or by 7or by 8 or by 9 or by 10 patients. Only case we are not interested in when none of the patients has side effect. So the required probability will be

P (Side effects occurs in a random sample of 10 patients taking the drug)
= P (x=1) + P (x=2) + P (x=3) + … + P (x=9) + P (x=10)

If we refer the properties of probability functions, sum of probabilities is 1 i.e.,
$\Sigma P(x) = 1$

In this experiment, X can take values as 0, 1, 2, 3, 4, 5, 6, 7, 8, 9, and 10, hence
$\Sigma P(x)$ = P (x=0) +P (x=1) + P (x=2) + P (x=3) + … + P (x=9) + P (x=10)
$\Rightarrow$ P (x=0) +P (x=1) + P (x=2) + P (x=3) + … + P (x=9) + P (x=10) = 1
$\Rightarrow$ P (x=1) + P (x=2) + P (x=3) + … + P (x=9) + P (x=10) = 1 - P (x=0)

So the desired probability
P (Side effects occurs in a random sample of 10 patients taking the drug)
= P (x=1) + P (x=2) + P (x=3) + … + P (x=9) + P (x=10)
= 1 - P (x=0)
= 1 – P (No person with side effect in 10 persons)
= $1 - {}^{10}C_0 p^0 q^{3-0}$
= $1 - 1*1*(0.97)^{10}$
= 1 – 0.7374
= 0.2626

Thus the chances that at least one person will have side effects are 26.26%.

Example: There are 3% chances that a batch of capsules will be declared as defective. What is the probability that none of the 5 batches will be defective? What is the probability that at least 3 of the 5 batches will be defective?

Solution:
- There are only two possible outcomes: either batch will be defective or not. Also these outcomes can not happen together.
- Having defects in one batch is independent of having defects in another batch.

As both the assumptions are satisfied in this situation, we can assume Binomial distribution and calculate the desired probabilities as follows:
Trial: One batch's checking
Experiment: Consists of 5 batch's checking, hence n =5
X = Number of batch having defects. It can be 0, 1, 2, 3, 4 or 5
p = Probability of having defects in one trial = 3/100 = 0.03

q = P (Not having defects in one trial)
    = 1 - P (Having defects in one trial) = $1 - 0.03 = 0.97$

We are interested in calculating the
(i) Probability that none of the 5 batches will be defective, i.e.,
P (No defective batch is found in 5 batches to be checked)
= P (x=0)
= ${}^5C_0 p^0 q^{5-0}$
= $1*1*(0.97)^5$
= 0.8587

(ii) Probability that at least 3 of the 5 batches will be defective
P (At least 3 of the 5 batches are found to be defective)
= P (x=3) + P (x=4) + P (x=5)
= ${}^5C_3 p^3 q^{5-3} + {}^5C_4 p^4 q^{5-4} + {}^5C_5 p^5 q^{5-5}$
= $10*0.03^3*0.97^2 + 5*0.03^4*0.97^1 + 1*0.03^5*0.97^0$
= 0.000254+0.000003933834+0.0000000243
= 0.000258

Thus the chances that none of the 5 batches will be defective are 85.87% and at least 3 of the 5 batches will be defective are 0.03%.

Example: Consider five mice from the same litter, all suffering from vitamin A deficiency. They are fed a certain dose of carrots. The positive reaction means recovery from the disease. Assume that the probability of recovery is 0.73. What is the probability that exactly or at least three of the five mice recover.

Solution: If mice are fed a certain dose of carrots,
- There will be only two outcomes, either mice will recover or mice will not recover. Also these outcomes can not happen together.
- Recovery of one mouse is independent of Recovery of another mouse.

As both the assumptions are satisfied in this situation, we can assume Binomial distribution and calculate the desired probabilities as follows:
Trial: 1 mouse taking a certain dose of carrots
Experiment: Consists of 5 mice taking a certain dose of carrots, hence n =5
X = Number of mice recovered after fed a certain dose of carrots i.e. it can be 0, 1, 2, 3, 4 or 5
p = Probability of recovery in one trial = P (Having recovery in one trial) = 0.73
q = P (Not having recovery in one trial)
    = 1 - P (Having recovery in one trial) = $1 - 0.73 = 0.27$

We are interested in calculating the probability of exactly or at least three of the five mice recover.
P (exactly three of the five mice recover taking a certain dose of carrots)

$= P(x=3)$
$= {}^5C_3 p^3 q^{5-3}$
$= 10*0.73^3*0.27^2$
$= 0.2836$

P (at least three of the five mice recover taking a certain dose of carrots)
$= P(x=3) + P(x=4) + P(x=5)$
$= {}^5C_3 p^3 q^{5-3} + {}^5C_4 p^4 q^{5-4} + {}^5C_5 p^5 q^{5-5}$
$= 10*0.73^3*0.27^2 + 5*0.73^4*0.27^1 + 1*0.73^5*0.27^0$
$= 0.2836 + 0.3834 + 0.2073$
$= 0.8743$

So probability of exactly or at least three of the five mice recover are 0.2836 and 0.8743 i.e. there are 28.36 % chances that exactly three of the five mice and 87.43% chances that at least three of the five mice will recover after taking a certain dose of carrots.

Example: 30% of patients afflicted with a certain disease die from it. What is the probability that none will die in a group of 5? What is the probability that at least 3 will die in a group of 5?

Solution: In this example,
- Outcomes are either person afflicted with a certain disease will survive or die. Also these two outcomes can not occur together
- "A person afflicted with that disease will die or survive" is independent of "another person afflicted with that disease will die or survive".

As both the assumptions of Binomial distributions are satisfied, we can assume Binomial distribution and calculate the desired probabilities as follows:

Trial: A person afflicted with a certain disease
Experiment: Consists of 5 patients, hence n =5
X = Number of success i.e. Number of persons who die after afflicting with a certain disease, hence it can be 0, 1, 2, 3, 4 or 5
p = Probability of success in one trial = P (death in one trial) = 0.30
q = 1-p = P (survival in one trial) = 1 - 0.30 = 0.70

P (None will die in a group of five i.e. X=0)
$= {}^5C_0 p^0 q^{5-0} = 0.1681$

P (at least 3 will die in a group of 5 i.e. X=3, 4, 5)
= P (3 will die in a group of 5) + P (4 will die in a group of 5) + P (5 will die in a group of 5)
$= {}^5C_3 p^3 q^{5-3} + {}^5C_4 p^4 q^{5-4} + {}^5C_5 p^5 q^{5-5}$

= 0.02835+0.1323+0.00243
= 0.16308

Thus the chances that all persons will survive in a group of 5 is 16.81%, at least 3 will die is 16.31%.

Poisson distribution
If in a Binomial distribution,
- n, the number of trials, is very large (tends to ∞)
- p, the probability of success in one trial is very small (tends to 0)
- mean $= \lambda = np$ is of moderate size, (0 to 10)

The Binomial distribution function is approximated by the Poisson distribution function given by

$$P(X = x) = \begin{cases} \dfrac{e^{-\lambda}\lambda^x}{x!} \ldots x = 0,1,2,\ldots,\infty, \lambda > 0 \\ 0 \ldots\ldots\text{otherwise} \end{cases}$$

The practical implication of this result is that in case of above 3 conditions, the Binomial probabilities may be well approximated by the much simpler Poisson probabilities which are given by only one parameter λ. λ is also the mean and variance of Poisson variable X.

To understand the concept, consider the following examples:

Example: If the probability that an individual suffers a bad reaction from injection of a given serum is 0.001, determine the probability that out of 2000 individuals, (a) exactly 3, (b) more than 2 individuals will suffer a bad reaction.

Solution: In this experiment,
If an individual is taking injection,
- There are only two possible outcomes: either he/she will suffer a bad reaction or not. Also these outcomes can not happen together.
- One person suffers a bad reaction is independent of another person suffers a bad reaction.

As both the assumptions are satisfied in this situation, we can assume Binomial distribution and calculate the desired probabilities as follows:

Trial: One individual gets an injection of a given serum
Experiment: Consists of 2000 individuals getting injection of a given serum, i.e., n = 2000
X = Number of individuals suffering a bad reaction, so it can be 0, 1, 2, ... or 2000

p = P (An individual suffers a bad reaction from injection of a given serum)
= 0.001
q = 1 – p = 1 – 0.001 = 0.999

According to Binomial distribution
Probability that out of 2000, exactly 3 will suffer a bad reaction

$$= P(X = 3) = {}^{2000}C_3 (0.001)^3 (0.999)^{1997}$$

$$= \frac{2000 * 1999 * 1998}{6} (0.001)^3 (0.999)^{1997}$$

$$= 0.1805$$

Here
- n=2000 which is large
- p= 0.001 which is very small
- λ = np = 2000*0.001 = 2 which is of moderate size

Hence Poisson distribution can be applied for calculation of desired probabilities
as follows:
P (Out of 2000, exactly 3 will suffer a bad reaction)

$$P(X = 3) = \frac{e^{-\lambda} \lambda^x}{x!} = \frac{e^{-2} 2^3}{3!} = 0.1804$$

It is clear that both the probabilities are very close. Hence it is easier to apply
Poisson distribution instead of Binomial distribution if all the 3 conditions are
satisfied.
So now we calculate Probability that out of 2000, more than 2 individuals will
suffer a bad reaction using Poisson distribution.
P(out of 2000, more than 2 individuals will suffer a bad reaction)
= P(X > 2)

$$= \sum_{x=3}^{2000} P(X = x)$$

On applying probability function's property $\sum_{x=0}^{2000} P(X = x) = 1$,

$$\sum_{x=3}^{2000} P(X = x)$$

$$= 1 - P(X = 0) - P(X = 1) - P(X = 2)$$

$$= 1 - \left(\frac{e^{-2} 2^0}{0!} + \frac{e^{-2} 2^1}{1!} + \frac{e^{-2} 2^2}{2!} \right)$$

$$= 1 - e^{-2} \left(\frac{2^0}{0!} + \frac{2^1}{1!} + \frac{2^2}{2!} \right) = 1 - e^{-2}(1 + 2 + 2)$$

$$= 1 - 5/e^2$$

$$= 0.323$$

Thus the chances that out of 2000 individuals, exactly 3 individuals will suffer a bad reaction is 18.04% and more than 2 will suffer a bad reaction is 32.3% based on the given data.

Example: Assume that a drug causes a serious side effect at a rate of three patients out of one hundred. What is the probability that the side effects occurs in a random sample of ten patients taking the drug?

Solution:
n = 10
X= Number of persons having serious side effects due to a drug i.e. it can be 0, 1, 2, 3, 4, 5, 6, 7, 8, 9 or 10
p = P(Drug causes a serious side effect) = 0.03
λ = np = 10*0.03= 0.3
As p is small and λ is of moderate size, we calculate the probability using Poisson distribution.

We wish to calculate the probability that the side effects occur in a random sample of ten patients taking the drug. This means our interest is in calculating the probability of having side effect either by 1 or by 2 or by 3 or by 4 or by 5 or by 6 or by 7or by 8 or by 9 or by 10 patients. Only case we are not interested in when none of the patients has side effect. So the required probability will be

P (Side effects occurs in a random sample of 10 patients taking the drug)

$$= P (x=1) + P (x=2) + P (x=3) + \dots + P (x=9) + P (x=10) = \sum_{x=1}^{10} P(X = x)$$

(On applying probability function's property i.e., $\sum_{x=0}^{10} P(X = x) = 1$)

$$= 1 - P (x=0)$$

$$= 1 - \frac{e^{-\lambda}\lambda^{x}}{x!}$$

$$= 1 - \frac{e^{-0.3}0.3^{0}}{0!}$$

$$= 1 - \frac{e^{-0.3}*1}{1}$$

$$= 1 - 0.7408$$

$$= 0.2592$$

This probability was 0.2626 when distribution was assumed as Binomial distribution (refer example on page no. 6.6-6.7). This infers that even **when n is as large as 10**, Poisson distribution approximates the Binomial distribution.

Example: 10% of the products produced in a certain manufacturing process turn out to be defective. Find the probability that in a sample of 10 products chosen at random, exactly two will be defective by using (a) the Binomial distribution, (b) Poisson approximation to Binomial distribution?

Solution:
n = 10
X= Number of defective products and hence can assume values as 0, 1, 2, 3, 4, 5, 6, 7, 8, 9 or 10
p = P(Product being defective in a lot) = 10/ 100 = 0.1 (as 10% products are defective)
q = 1-p = 1 – 0.1 = 0.9

$$\lambda = np = 10*0.1 = 1$$

Under Binomial distribution
P (Number of exactly two defective products)
$$= P(X = 2) = {}^{10}C_2 (0.1)^2 (0.9)^8$$

$$= \frac{10*9}{2} * 0.0043$$

$$= 0.1937$$

Under Poisson approximation to Binomial distribution
P (Number of exactly two defective products)
= P (X=2)

$$= \frac{e^{-\lambda}\lambda^x}{x!}$$

$$= \frac{e^{-1}1^2}{2!}$$

$$= \frac{e^{-1}}{2!}$$

$$= 0.1839$$

This example illustrates that when p is not very small and n is also not large, Poison distribution should not be applied in place of Binomial distribution.

Poisson distribution, besides providing an approximation to the Binomial distribution plays an important role in its own right.

Suppose a random event satisfies the following three conditions:
- Events occurring in one time interval are independent of those occurring in any other mutually exclusive interval
- The probability of events occurring is proportional to the length of the time interval.
- The probability of two or more events occurring in an infinitesimally small interval is negligible

Then the distribution of the random variable X where X= number of events occurring in a specified time interval of length t, is given by a Poisson distribution with parameter λt, where λ = mean number of occurrence per unit time interval. Its distribution function is given by

$$P(X = x) = \begin{cases} \dfrac{e^{-\lambda t}(\lambda t)^x}{x!} \ldots\ldots x = 0,1,2,\ldots,\infty, \lambda > 0 \\ 0 \ldots\ldots\ldots \text{otherwise} \end{cases}$$

In this case, the mean and variance of Poisson distribution are λt and λt.

Poisson distribution is applicable to time intervals as well as to sections of space. If instead of intervals of length t we have domains of area or volume, the same argument applies to the distribution of points.

Example: *Radioactive disintegrations:* A radioactive substance emits α particles, the number of particles reaching a given portion of space during time t is the best known example of random event obeying the Poisson law. Of course, the substance continues to decay, and in the long run the density of α particles will decline.

Example: *Chromosome interchanges in cells:* Irradiation by X-ray procedures produces certain processes in organic cells which we call chromosome interchanges. As long as radiation continues, the probability of such interchanges remains constant, and according to theory, the number Nk of cells with exactly k interchanges should follow a Poisson distribution. The theory is also able to predict the dependence of the parameter λ on the intensity of radiation, the temperature, etc.

Example: When counting red blood cells, a square grid is used, over which a drop of blood is evenly distributed. Under the microscope an average of 8 erythrocytes are observed per single square. What is the probability that exactly 5 erythrocytes are found in one square?

Solution:
 (i)   Number of erythrocytes in one square is independent of number of erythrocytes in another square
 (ii) More number of erythrocytes will be found if size of squares is big
 (iii) The probability of finding two or more erythrocytes in an infinitesimally small square is negligible

As all the assumptions are satisfied, Poison distribution can be applied. Here
$\lambda = 8$ erythrocytes / single square
t = 1 square
Hence Mean = $\lambda t = 8$

X = number of erythrocytes

P (Exactly 5 erythrocytes are in one square)
= P (x = 5)

$$= \frac{e^{-\lambda t}\lambda t^{x}}{x!} = \frac{e^{-8}8^{5}}{5!} = \frac{0.000335 * 32768}{120} = \frac{10.99244}{120} = 0.0916$$

The probability that exactly 5 erythrocytes are found in one square is 0.0916, i.e. there are 9.16% chances that exactly 5 erythrocytes are found in one square.

Example: In an experiment with radioactive tracer, 4.4 scintillations per second were observed on the average. Find the probability that at least one scintillation occurs in a given time interval of length, i.e., per sec.

Solution:
λ = 4.4 scintillations / sec
t = 1 sec
Mean = λt = 4.4
X = number of scintillation

P(At least one scintillation occurs)
= P (X $\geq$ 1)
= P (x = 1) + P (x = 2) + P (x = 3) + ...
On applying probability function property that sum of all probabilities is 1, we get
= 1 - P (x=0)

$$= 1 - \frac{e^{-\lambda t}(\lambda t)^{x}}{x!}$$

$$= 1 - \frac{e^{-4.4}4.4^{0}}{0!}$$

$$= 1 - e^{-4.4}$$
$$= 0.9877$$

Probability that at least one scintillation occurs per sec is 0.9877.

Example: As a routine quality control, 100 cartons, each containing 50 bottles of IV fluids were inspected. The results were tabulated as follows:

Number of bottles found defective	Number of cartons
0	4
1	14
2	23

3	23
4	18
5	9
6	9

Find the probability of having a carton with no bottle defective.

Solution: Here data is given in the form of frequency distribution, hence average number of bottles found defective per carton will be calculated using formula for frequency distribution.

Average number of bottles found defective per carton $= \bar{x} = \dfrac{\sum f_i x_i}{\sum f_i} = \dfrac{300}{100} = 3.0$

$\lambda = 3$ defective bottles / carton
$t = 1$ carton
$\lambda t = \text{Mean} = 3$
$X = \text{number of defective bottles}$

P (No defective bottle in a carton)
$= P\,(X = 0)$
$= \dfrac{e^{-\lambda t}(\lambda t)^x}{x!}$
$= \dfrac{e^{-3}\,3^0}{0!}$
$= 0.049797$

i.e., there are 5% chances that there will be no defective bottle in a carton.

Example: The distribution of cells in squares of haemocytometer has a Poisson distribution. Following is the data of number of cells per square obtained from 100 squares.

Number of cells per square	0	1	2	3	4	5	6	7	>7
Number of squares	7	11	20	25	14	13	7	3	0

Calculate the expected number of squares (frequencies) as per Poisson distribution for number of cells per square.
Solution:
Data is given in the form of frequency distribution, so

Average cells per square $= \bar{x} = \dfrac{\sum f_j x_j}{\sum f_j} = \dfrac{310}{100} = 3.1$

Hence $\lambda = 3.1$

X is the number of cells in squares of haemocytometer and it takes values 0, 1, 2, 3, 4, 5, 6, 7 or >7.

Poisson probabilities for each value of X are calculated using the formula $\dfrac{e^{-\lambda}\lambda^{x}}{x!}$.

To calculate the expected / predicted number of squares (frequencies) as per Poisson distribution, probabilities are multiplied by the total number of squares i.e., 100. Expected frequencies are as follows:

Table: Poisson probabilities and expected frequencies for number of cells per square

Number of cells per square (X)	Number of squares (Observed frequency)	Poisson probabilities (p)	Expected number of squares as per Poisson distribution	Expected number of squares (frequencies) (After rounding off)
0	7	0.0450	4.50492	5
1	11	0.1397	13.96525	14
2	20	0.2165	21.64614	22
3	25	0.2237	22.36768	22
4	14	0.1733	17.33495	17
5	13	0.1075	10.74767	11
6	7	0.0555	5.552963	6
7	3	0.0246	2.459169	2
>7	0	0.0095	0.952928	1

Following figure graphically represents the observed and expected (predicted) frequencies for number of cells per square.

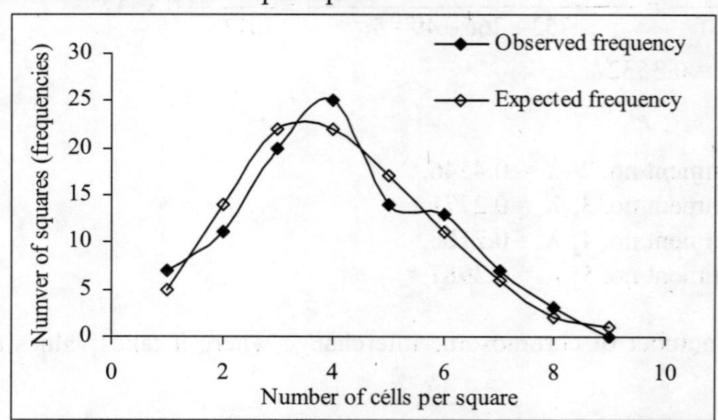

Figure: Pictorial presentation of observed and predicted number of squares (frequencies) for number of cells per square

The table values and figure show that there is a concordance between observed and predicted frequencies. This confirms that the number of cells in squares of haemocytometer follows a Poisson distribution.

Example: Number of cells with 0, 1, 2, or $\geq$ 3 chromosome interchanges induced by X-ray irradiation are obtained for 5 experiments as follows.

	Number of cells			
	Number of chromosome interchanges			
Experiment No.	**0**	**1**	**2**	**$\geq$ 3**
1	753	266	49	5
2	434	195	44	9
3	280	75	12	1
4	639	141	13	0
5	793	339	62	5

As number of chromosome interchange follows a Poisson distribution, calculate expected number of cells with chromosome interchanges.

Solution:
Here X is the number of chromosome interchanges which assumes values as 0, 1, 2, or $\geq$ 3 and the corresponding number of cells with a specific number of chromosome interchanges (frequencies) are given row wise for each experiment.

As data is given in the form of frequency distribution, for experiment no.1,
Average number of chromosome interchange

$$= \bar{x} = \frac{\sum f_j x_j}{\sum f_j} = \frac{0*753 + 1*266 + 2*49 + 3*5}{753 + 266 + 49 + 5} = \frac{379}{1073} = 0.3532$$

Hence $\lambda = 0.3532$.

Similarly,
For experiment no. 2, $\lambda = 0.4546$.
For experiment no. 3, $\lambda = 0.2771$.
For experiment no. 4, $\lambda = 0.2106$.
For experiment no. 5, $\lambda = 0.3987$.

X is the number of chromosome interchange where it takes values 0, 1, 2 and $\geq$ 3.

Poisson probabilities for variable X are calculated using the formula $\dfrac{e^{-\lambda}\lambda^x}{x!}$.

These probabilities are multiplied by the total frequencies to calculate the expected number of cells with chromosome interchanges, i.e.
For experiment no. 1, probabilities are multiplied by 1073.
For experiment no. 2, probabilities are multiplied by 682.
For experiment no. 3, probabilities are multiplied by 368.
For experiment no. 4, probabilities are multiplied by 793.
For experiment no. 5, probabilities are multiplied by 1199.

Expected frequencies are as follows:

Experiment No.		Cells with X interchanges				N	λ
		0	1	2	≥ 3		
1	Observed frequency	753	266	49	5	1073	0.3532
	Expected frequency	752	267	47	6		
2	Observed frequency	434	195	44	9	682	0.4546
	Expected frequency	432	197	45	8		
3	Observed frequency	280	75	12	1	368	0.2771
	Expected frequency	279	77	11	1		
4	Observed frequency	639	141	13	0	793	0.2106
	Expected frequency	642	135	142	1		
5	Observed frequency	793	339	62	5	1199	0.3987
	Expected frequency	805	321	64	9		

Table values reveal that observed and expected number of cells (frequencies) with specific number of chromosome interchange are very close and hence number of chromosome interchange follows a Poisson distribution.

Example: To study the effect of high humidity on capsules packed in glassing-poly Strips, 200 strips containing 10 capsules each were kept in an environment of 95% RH for a period of 30 days. At the end of 30 days, capsules were de-stripped to find out which ones had become sticky due to moisture pick-up. Observations are as follows:

Number of sticky capsules in a strip (X)	Number of strips (frequency (f))
0	122
1	60
2	15
3	2
4	1

Test whether number of sticky capsules follows a Poisson distribution.

Solution:
Data is given in the form of frequency distribution, hence

Average number of sticky capsules per strip $= \bar{x} = \dfrac{\sum f_j x_j}{\sum f_j} = \dfrac{100}{200} = 0.5$

Hence $\lambda = 0.5$

X is the number of sticky capsules where it takes values 0, 1, 2, 3 or 4.
Poisson probabilities of number of sticky capsules are calculated using the

formula $\dfrac{e^{-\lambda}\lambda^x}{x!}$.

To calculate the expected frequencies (NP(X=x), where X=0, 1, 2, 3 or 4) as per

Poisson distribution, probabilities ($\dfrac{e^{-\lambda}\lambda^x}{x!}$, for X =0, 1, 2, 3 and 4) are multiplied

by the total frequencies i.e. 200. Hence $NP(X = x) = N\dfrac{e^{-\lambda}(\lambda)^x}{x!}$

So expected frequencies are as follows:

$P(0) = \dfrac{e^{-0.5}0.5^0}{0!} = 0.607$

$\therefore NP(0) = 200 \times 0.607 = 121$

$NP(1) = N\dfrac{e^{-0.5}(0.5)^1}{1!} = 200 \times 0.30 = 60$

$NP(2) = N\dfrac{e^{-0.5}(0.5)^2}{2!} = 200 \times 0.076 = 15$

$NP(3) = N\dfrac{e^{-0.5}(0.5)^3}{3!} = 200 \times 0.013 = 2.6 \approx 3$

$NP(4) = N\dfrac{e^{-0.5}(0.5)^4}{3!} = 200 \times 0.003 = 0.63 \approx 1$

Table: Observed and expected number of strips having 0, 1, 2, 3 or 4 sticky capsules

Number of sticky capsules in a strip	Observed number of strips	Expected number of strips
0	122	121
1	60	60
2	15	15
3	2	3
4	1	1

As observed and expected number of strips are almost same, it infers that number of sticky capsules follows a Poisson distribution.

Continuous Distribution

Continuous variable X can assume any value in an interval, i.e., for each value of x we have a value of P(X=x) which is denoted by a continuous function f(x) in continuous variable case. So the histogram will converge to a continuous curve f(x).

Some examples of f(x) are as follows:

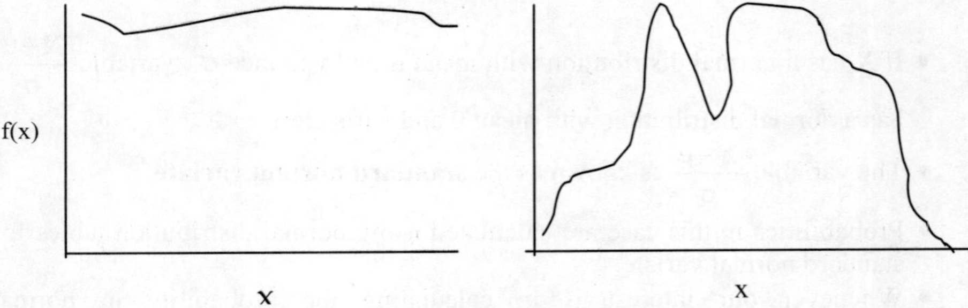

Figure: Some continuous distributions

f(x) is known as the density function of random variable X.

In line with discrete variable case, f(x) also satisfies the following properties:
1. $f(x) \geq 0$ for all values of X
2. $\int f(x) = 1$ (instead of $\sum f(x)$, $\int f(x)$ is used as X is a continuous variable)

Normal distribution

The normal distribution is defined by the probability density function

$$f(x) = \frac{1}{\sigma\sqrt{2\pi}} e^{-\frac{(x-\mu)^2}{2\sigma^2}} \dots\dots\dots\dots\dots\dots\dots\dots -\infty < x < \infty$$

$$= 0 \dots\dots\dots\dots\dots\dots\dots\dots\dots\dots\dots\text{Otherwise},$$

where μ = Mean of the distribution

σ = Standard deviation of the distribution.

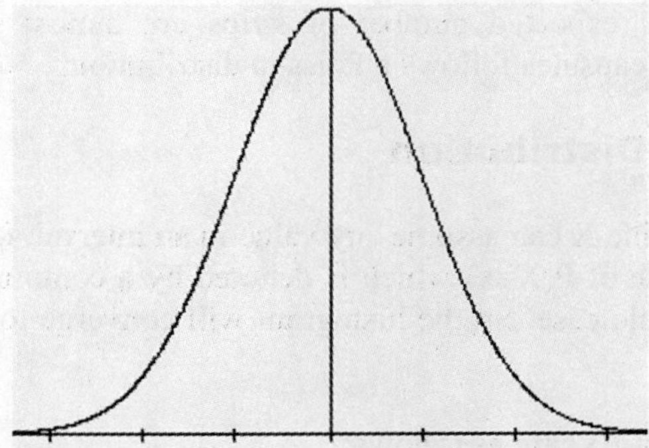

Figue: Pictorial representation of normal distribution

- If X has a normal distribution with mean μ and variance σ^2, variable $\dfrac{X-\mu}{\sigma}$ has a normal distribution with mean 0 and variance 1.

- The variable $\dfrac{X-\mu}{\sigma}$ is known as the **standard normal variate**.

- Probabilities in this case are calculated using normal distribution tables for standard normal variate.

- Whenever our interest is in calculating the probability in normal distribution case, variable has to be transformed into a standard normal variate because tables are easily available for standard normal variate in many books.

How to look at the tables

- If our objective is to find out the probability of X<x, transform the variable into standard normal variate. We will get a number. The digit before decimal of the number together with first digit after decimal of the number are given in rows of the normal distribution tables, and second digit after decimal is mentioned in columns. We have to see the cross section of these. For example, if the value of standard normal variate is 1.96, 1.9 should be seen in row and 6 in column and then the probability is the value written in intersection of these.

- If objective is to find out the value of the standard normal variate based on the given probability of X<x, find out where this probability value lie in the table. Then corresponding values in row and column should be seen. First write the row value (which has decimal) and then the column value at the second decimal place. This is the value of the standard normal variate.

We will see this through an example.

Example: A population of marine gastropods has shell lengths which are normally distributed with mean 7 mm and variance 2 mm^2. What proportion of the population will have a shell length between 5 and 9 mm?

Solution:
Let X denote the shell length,
Mean of $X = \mu = 7$
Variance of $X = \sigma^2 = 2$

Hence the standard normal variate $= Z = \dfrac{X-\mu}{\sigma} = \dfrac{X-7}{\sqrt{2}}$

We want to calculate the probability of X between length 5 and 9 mm.

$P(5<X<9)$

$= P(\dfrac{5-7}{\sqrt{2}} < \dfrac{X-7}{\sqrt{2}} < \dfrac{9-7}{\sqrt{2}})$ (To make X a standard normal variate)

$= P(-\sqrt{2} < Z < \sqrt{2})$

$= P(-1.4142 < Z < 1.4142)$

$= 0.9207 - 0.0793$ (On using the tables of standard normal variate)

$= 0.8414$

Hence approximately 84% of the population will have a shell length in the range 5 to 9 mm.

Example: If the time taken for an eyelid to close after a foreign particle strikes the surface of the eyeball (closing time) is normally distributed with mean 5 microseconds, calculate the standard deviation in order that the closing time is less than 4 microseconds with probability 0.3.

Solution:
Let T be the closing time and its variance is σ,
$\mu =$ Mean of $T = 5$

It is given that $P\,(T < 4) = 0.3$

To make "T" a standard normal variate, subtract 5 and divide by standard deviation σ, this gives

$p[\dfrac{T-5}{\sigma} < \dfrac{4-5}{\sigma}] = 0.3$

$p[Z < \dfrac{-1}{\sigma}] = 0.3$

From the normal distribution table, the value of Z which gives the probability as 0.3 is -0.53.

$\Rightarrow -0.53 = -1/\sigma$

$\Rightarrow \sigma = 1.88$

Hence the standard deviation for closing time is 1.88 microsecond^2.

> **Also for normal distributions,**
> μ ± **1 standard deviation limits contain 68.3% of the data.**
> μ ± **2 standard deviation limits contain 95.4% of the data.**
> u ± **3 standard deviation limits contain 99.7% of the data.**

Example: Consider the vincamine production in multiple shoot culture derived from hairy roots of vinca minor. Following are the vincamine content (%dwb) in 20 regenerated shoots.

0.53	0.50	0.54	0.48	0.57	0.55	0.51	0.52	0.49	0.50
0.50	0.54	0.56	0.51	0.53	0.53	0.55	0.52	0.51	0.52

Average vincamine production in regenerated shoots and its standard deviation (Refer page 4.23 for calculations)

$$\text{Average vincamine production} = \frac{10.46}{20} = 0.523$$

sd = 0.0249

If we look at the data, μ ± 1 standard deviation i.e. 0.523 ± 0.0249 = (0.4981, 0.5479) limits contain 14 observations (70% of the data).

Similarly μ ± 2 standard deviation i.e. 0.523 ± 2*0.0249 = (0.4732, 0.5728) limits contain 20 observations (100% of the data).

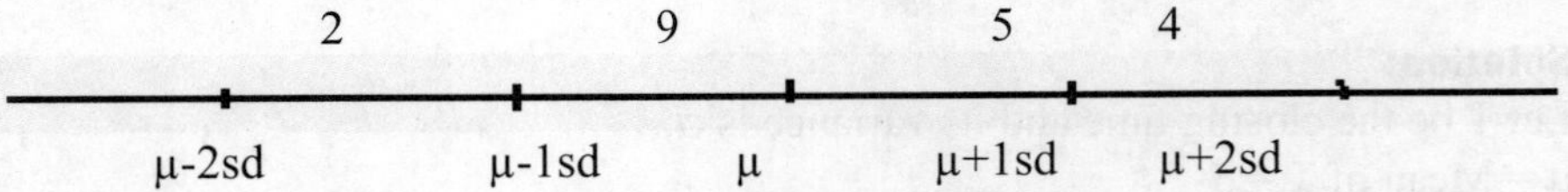

Figure: Pictorial representation of distribution of vincamine production in multiple shoot culture

Example: Consider the number of shoots regenerated from leaf callus of Sempervivum tectorum. Following table gives the number of shoots regenerated per 10 gm leaf callus from 30 samples.

Sample No.	Number of shoots	Sample No.	Number of shoots	Sample No.	Number of shoots
1	20	11	20	21	18
2	18	12	19	22	22
3	19	13	18	23	24
4	23	14	21	24	16
5	22	15	19	25	20

6	19	16	20	26	25
7	17	17	20	27	15
8	20	18	21	28	22
9	19	19	19	29	23
10	21	20	19	30	21

Find out the distribution of number of shoots regenerated from leaf callus of Sempervivum tectorum.

Solution: Frequency table of number of shoots regenerated from leaf callus of Sempervivum tectorum using 30 samples is given below.

Table: Frequency distribution of number of shoots regenerated from leaf callus of Sempervivum tectorum (n=30)

Class Internal of number of shoots regenerated	Number of samples (f_j)
14-15	1
16-17	2
18-19	10
20-21	10
22-23	5
24-25	2

This data is presented pictorially using histogram.

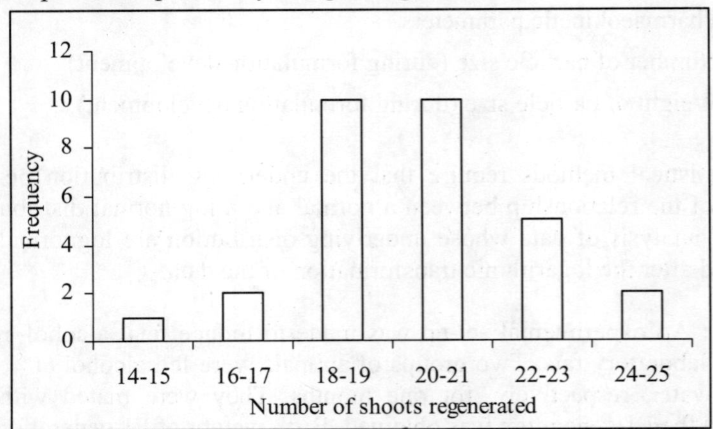

Figure: Histogram showing shoots regenerated per 10 gm from leaf callus of Sempervivum tectorum from 30 samples

Histogram shows that the distribution of number of shoots regenerated from leaf callus of Sempervivum tectorum is normal distribution.

Lognormal Distribution

- This distribution is closely related to the normal distribution. However, variable here can assume only non negative values
- The non-negative continuous variable X is said to have a log normal distribution if log X is normally distributed.
- The density function of a log normal variable is given by

$$f(x) = \frac{1}{x\sigma\sqrt{2\pi}} e^{-\frac{(\log x - \mu)^2}{2\sigma^2}}, x > 0$$

where μ = Mean of **logx**

σ = Standard deviation of **logx**.

i.e., μ and σ are calculated after data is log transformed.

If μ_x and σ_x^2 are the mean and variance befor log transformation, the μ_x and σ_x^2 of the lognormal variable X in terms of μ and σ^2 are

$$\mu_x = e^{\mu + \sigma^2/2}$$

$$\sigma_x^2 = (e^{\sigma^2} - 1)e^{2\mu + \sigma^2}$$

Log normal distribution is commonly used in many biological situations, for example,

- Weight of newborn babies
- Cholesterol level
- Pharmacokinetic parameters
- Number of particle size (during formulation development)
- Weight of particle size (during formulation development)

Many statistical methods require that the underlying distribution be normal. Because of the relationship between a normal and a log normal distribution, the statistical analysis of data whose underlying distribution are lognormal, can be performed after the logarithmic transformation of the data.

Example: An experimental set-up was made to induce fetal alcohol syndrome (FAS) in laboratory rats. Two groups of animals were fed alcohol (1% v/v) and distilled water, respectively, for one months. They were mated with control males and the F1 generation was obtained. Birth weight of F1 generation rats are given below in gms.

13.0	14.7	11.2	9.5	18.4
13.2	11.8	8.5	12.4	13.2
16.4	11.8	17.4	10.6	11.8
14.7	11.2	14.7	10.0	10.6

Find out the distribution of the data with parameters, i.e., mean and standard distribution.

Solution:

First prepare frequency distribution table of birth weight of F1 generation rats.

Table: Frequency distribution table of birth weight of F1 generation rats with other calculations required for mean and sd

Class interval of Birth weight	Number of rats (f_j)	Midpoint of class interval (x_j)	$x_j f_j$	$f_j x_j^2$
8.0-10.0	3	9.0	27.0	243
10.1-12.0	7	11.05	77.35	854.72
12.1-14.0	4	13.05	52.20	681.21
14.1-16.0	3	15.05	45.15	679.51
16.1-18.0	2	17.05	34.10	581.40
18.1-20.0	1	19.05	19.05	362.90
		Total	254.85	3402.74

Mean and sd of birth weight of F1 generation rats using the above frequency table

$$\bar{x} = \frac{\sum f_j x_j}{n} = \frac{254.85}{20} = 12.7425$$

$$s = \sqrt{\frac{1}{\sum_{i=1}^{n} f_i - 1}\left[\sum_{i=1}^{n} f_i x_i^2 - \frac{(\sum_{i=1}^{n} f_i x_i)^2}{\sum_{i=1}^{n} f_i}\right]} = \sqrt{\frac{1}{20-1}\left[3402.74 - \frac{(254.85)^2}{20}\right]}$$

$$= \sqrt{0.0526 \times 155.314} = sd = \sqrt{8.1744} = 2.8591$$

Graphical presentation of data is done by histogram.

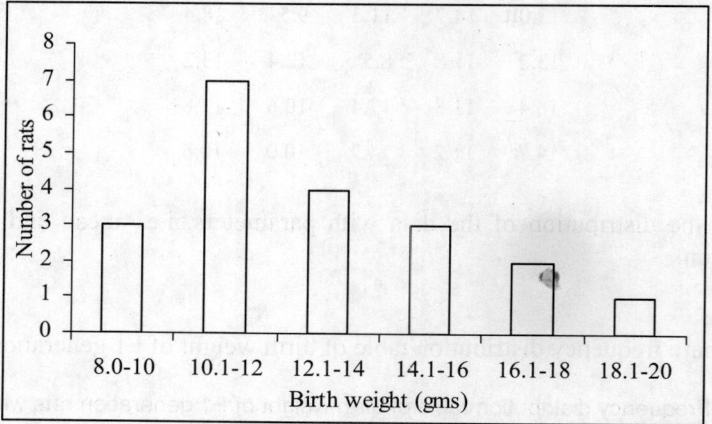

Figure: Histogram showing distribution of birth weight of F1 generation rats

It is clear from the histogram that data does not have a normal distribution.

Now if we assume log normal distribution of the birth weight, first we transform the data logarithmically as follows:

0.56	2.69	2.42	2.25	2.91
2.58	2.47	2.14	2.52	2.58
2.80	2.47	0.86	2.36	2.47
2.60	2.42	2.69	2.30	2.36

Table: Frequency distribution table of birth weight of F1 generation rats after logarithmic transformation with other calculations required for mean and sd

Class interval of Birth weight	Number of rats (f_j)	Midpoint of class interval (x_j)	$x_j f_j$	$f_j x_j^2$
2.0-2.15	1	2.075	2.075	4.30
2.16-2.30	2	2.23	4.46	9.94
2.31-2.45	4	2.38	9.52	22.65
2.46-2.60	7	2.53	17.71	44.80
2.61-2.75	3	2.68	8.04	21.54
2.76-2.90	2	2.83	5.66	16.01
2.91-3.05	1	2.97	2.97	8.85
Total	20		50.44	128.10

After logarthmic transformation

$$\overline{X} = \frac{\Sigma f_j x_j}{\Sigma f_j} = \frac{50.44}{20} = 2.522$$

$$s = \sqrt{\frac{1}{\sum\limits_{i=1}^{n} f_i - 1}\left[\sum\limits_{i=1}^{n} f_i x_i^2 - \frac{\left(\sum\limits_{i=1}^{n} f_i x_i\right)^2}{\sum\limits_{i=1}^{n} f_i}\right]} = \sqrt{\frac{1}{20-1}\left[128.10 - \frac{(50.44)^2}{20}\right]} = 0.2165$$

So mean and sd of original data will be

$$\mu = e^{\mu + o^2/2} = e^{2.522 + 0.21651^2/2} = 12.75$$

$$\sigma^2 = (e^{\sigma^2} - 1)e^{2\mu + \sigma^2} = (e^{0.2165^2} - 1)e^{2*2.522 + 0.2165^2} = 7.7972$$

$$\sigma = 2.7923$$

where $\overline{X}$ and s were used as estimate of μ and σ. These values are very close to the directly calculated mean and sd of original data (12.74 and 2.859, refer page 6.27).

Graphical presentation of birth weight of F1 generation rats assuming log normal distribution is as follows.

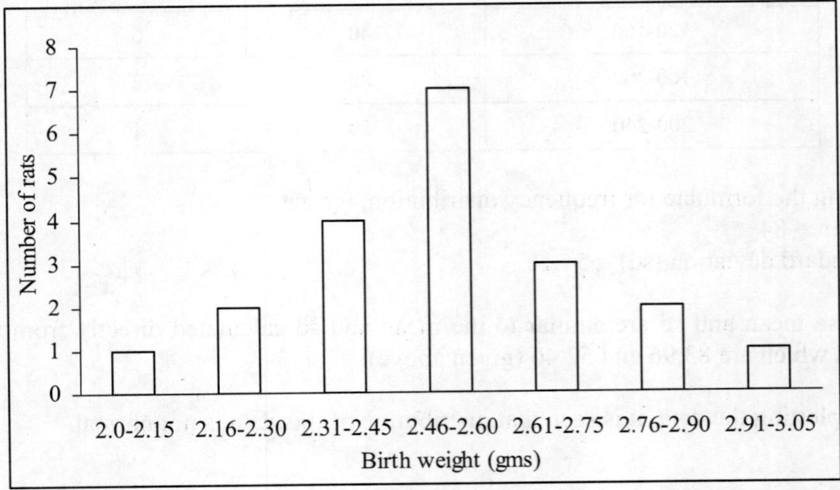

Figure: Histogram showing distribution of birth weight of F1 generation rats assuming log normal distribution i.e., after logarithmic transformation

Histograms show that assumption of log normal distribution of data is more appropriate in this case.

Example: An improved method for somatic plantlet production in hybrid larch (larix) was developed. Using this method, an experiment was conducted with 25 plants. Numbers of mature somatic embryos were obtained as follows:

58	137	69	61	42
66	83	38	29	25
188	32	83	162	102
19	51	130	77	153
47	74	43	49	221

Find out the distribution of the numbers of mature somatic embryos.

Solution:

Mean and sd of number of mature somatic embryos are calculated directly from the data and are as follows:

Mean = 83.96 , standard deviation (sd) = 52.46

Table: Frequency distribution of mature somatic embryos

Class interval of mature somatic embryos	Mid point of class interval	Number of plants (frequency)
0-40	20	4
40-80	60	12
80-120	100	3
120-160	140	3
160-200	180	2
200-240	220	1

Using the formulae for frequency distribution, we get

Mean = 84

Standard deviation (sd) = 54.16

These mean and sd are similar to the mean and sd calculated directly from the data which are 83.96 and 52.46 (given above).

On plotting the data, it is seen that distribution of this data is not normal.

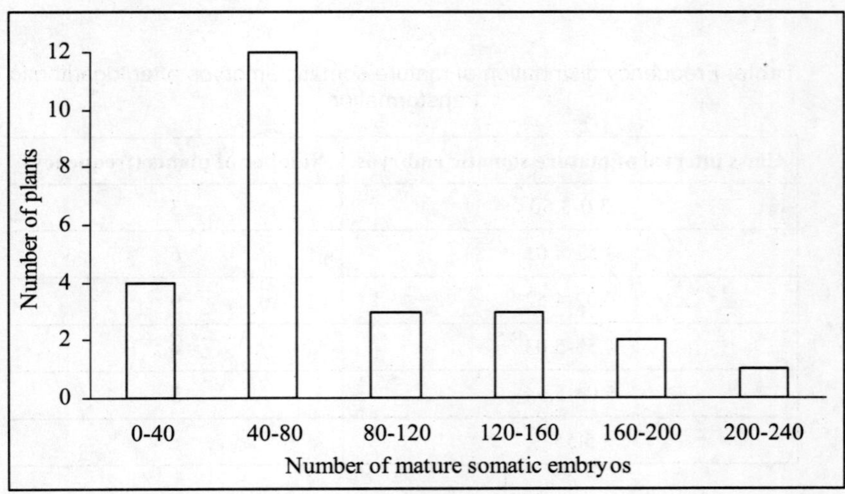

Figure: Histogram showing distribution of number of mature somatic embryos obtained from 25 plants

Hence we transform the data logarithmically as follows:

4.06	4.9	4.23	4.11	3.74
4.19	0.42	3.64	3.37	3.22
5.33	3.46	4.42	5.09	4.62
4.37	3.93	4.87	4.34	5.03
3.85	4.30	3.76	3.89	5.39

Mean and Standard deviation (sd) directly from the data after logarithmic transformation are:

Mean = 4.23

Standard deviation (sd) = 0.599

Hence mean and standard deviation of the original data will be

$$\text{Mean} = e^{\mu + \sigma^2/2} = e^{4.23 + 0.599^2/2} = 82.22$$

$$\text{Variance} = (e^{\sigma^2} - 1)e^{2\mu + \sigma^2} = (e^{0.599^2} - 1)e^{2*4.23 + 0.599^2} = 2917.76$$

$$\text{sd} = 54.02$$

where $\overline{X}$ and s were used as estimate of μ and σ. Which are similar to the mean and standard deviation calculated directly, i.e., mean = 83.96 , standard deviation (sd) = 52.46.

Now obtain frequency distribution of somatic embryos after logarithmic transformation as follows:

Table: Frequency distribution of mature somatic embryos after logarithmic transformation

Class interval of mature somatic embryos	Number of plants (frequency)
3.0-3.50	3
3.51-4.01	6
4.02-4.52	9
4.53-5.03	4
5.04-5.54	3
5.55-	

This data is again plotted as follows:

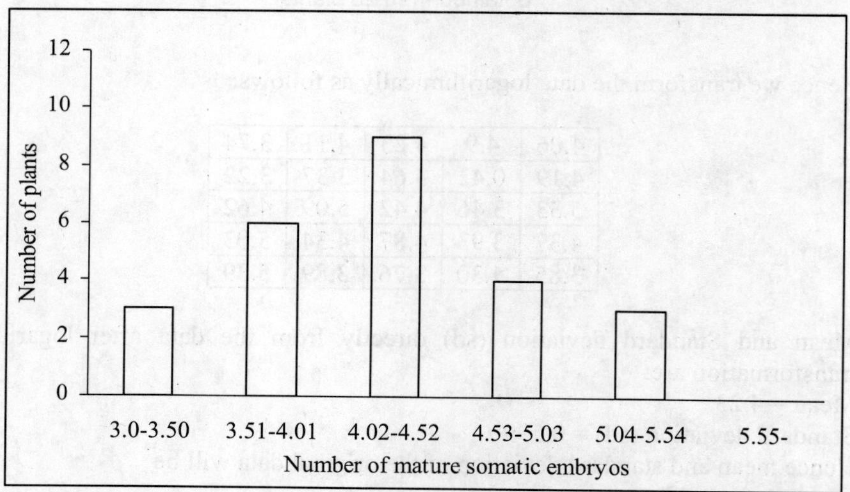

Figure: Histogram showing distribution of number of mature somatic embryos obtained in the 25 plants after logarithmic transformation

The histogram shows that distribution of number of mature somatic embryos has a lognormal distribution.

Example: During a bioequivalence study, the peak drug concentrations (ng/ml) for 20 volunteers are obtained and given below:

Volunteer No.	Peak drug concentration (ng/ml)	Volunteer No.	Peak drug concentration (ng/ml)
1	91	11	121.8
2	102.9	12	83.8
3	82.6	13	128.8
4	59.5	14	74.2
5	66.5	15	114.8
6	70	16	82.6
7	82.6	17	78.4
8	92.4	18	102.9
9	102.9	19	74.2
10	78.4	20	92.6

Find out the distribution of peak drug concentration.

Solution: After selecting class interval, the data may be written as follows.

Table: Frequency distribution of peak drug concentration obtained from 20 volunteers

Class interval of peak drug concentrations	Number of volunteers (frequency)
55-69	2
69-83	8
83-97	4
97-111	3
111-125	2
125-139	1

Graphical presentation of the data is as follows.

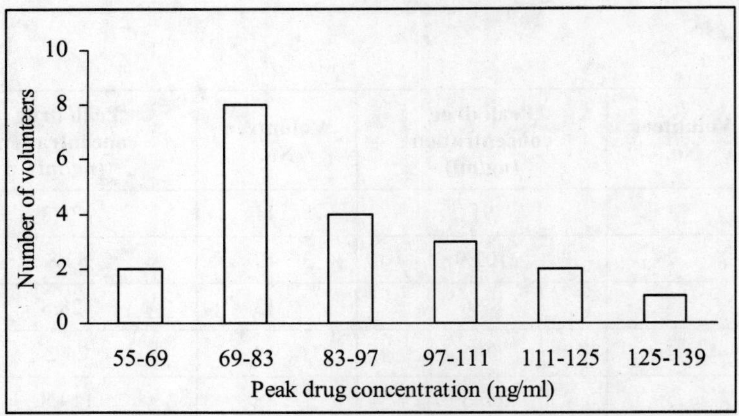

Figure: Histogram showing distribution of peak drug concentrations (ng/ml) for 20 volunteers

The above graph shows that data does not follow normal distribution. Hence the log values for above data are taken and are as below.

Table: Peak drug concentration after logarithmic transformation obtained from 20 volunteers

Volunteer No.	Peak drug concentration (ng/ml)	Volunteer No.	Peak drug concentration (ng/ml)
1	4.5108	11	4.8023
2	4.6337	12	4.4636
3	4.414	13	4.8552
4	4.0859	14	4.3067
5	4.1972	15	4.7431
6	4.2485	16	4.454
7	4.414	17	4.3618
8	4.5261	18	4.6337
9	4.6337	19	4.3067
10	4.3618	20	4.5261

To present the data graphically, we prepare the frequency table of the above data as follows:

Table: Frequency distribution of peak drug concentration after logarithmic transformation obtained from 20 volunteers

Class interval of peak drug concentrations	Number of volunteers (frequency) (f_i)	Mid point Class interval of peak drug concentrations (x_i)	$f_i x_i$	$f_i x_i^2$
4.05-4.20	2	4.125	8.25	34.03
4.20-4.35	3	4.275	12.82	54.83
4.35-4.50	6	4.425	26.55	117.48
4.50-4.65	6	4.575	27.45	125.58
4.65-4.80	2	4.725	9.45	44.65
4.80-4.95	1	4.875	4.87	23.76
		Total	89.4	400.33

Histogram of the log transformed peak drug concentration is given below.

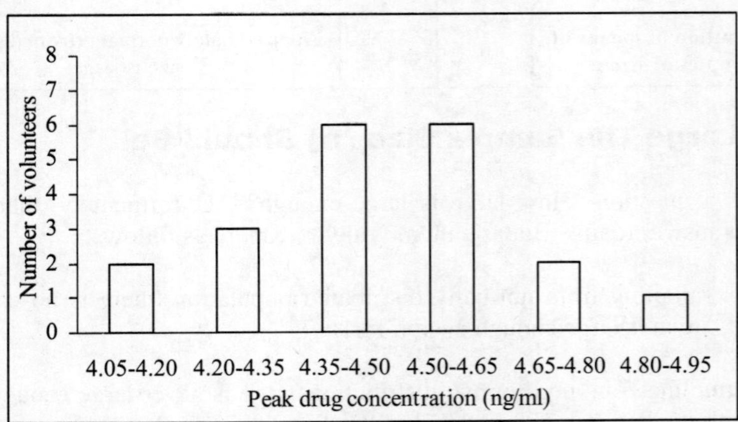

Figure: Histogram showing distribution of peak drug concentrations (ng/ml) for 20 volunteers after logarithmic transformation

This shows that data has a log normal distribution.

CENTRAL LIMIT THEOREM (CLT)

Let $x_1, x_2, \ldots\ldots, x_n$, be a random sample of size n from any distribution $f(x)$ with mean μ and variance σ^2, then

$\dfrac{\overline{X}_n - \mu}{\sigma/\sqrt{n}}$ has normal distribution with mean 0 and variance 1 when sample size is large i.e., $n \to \infty$.

This is also represented as

$$\frac{\overline{X}_n - \mu}{\sigma/\sqrt{n}} \to \quad N(0, 1) \text{ as } n \to \infty$$

where $\overline{x}_n$ = Sample mean = $\dfrac{x_1 + x_2 + \ldots\ldots + x_n}{n}$

The remarkable thing about this theorem is that it holds true whatever the distribution of X, provided it has a finite variance.

Table: Summary of properties of the sampling distribution of means compared with the underlying population distribution

	Mean	SD	Shape
Distribution of Population	μ	σ	Any
Distribution of means of samples of size n	μ	$\sigma/\sqrt{n}$	Approximately normal provided sample size is large

How Large The Sample Size (n) Should Be

There is a question "How large is large enough?". Unfortunately, there is no clear cut answer to this. But as a thumb rule we can go as follows:

If one is sampling from normally distributed population, there is no cause for concern. Any n is large enough even if n=1.

When sampling from non-normal distributions, n=1 is never large enough, since the sampling distribution of means of samples of size 1 merely reduces the population distributions.

In general, n at which the sampling distribution of means tends to be approximately normal in shape depends on the extent to which the underlying population distribution departs from a normal distribution shape.

Importance of CLT

Finding out the theoretical distribution of a variable directly from the data is not easy due to the following reasons:

- The sample size may not be large enough to determine the theoretical distribution with much precision.
- Even though we may have data from many experiments, we can not pool the data of all experiments and assume that it is from one experiment to determine the theoretical distribution. This is because (i) all the samples might not have taken from population randomly (statistical procedures typically assume that the samples are obtained in a random fashion to reach valid conclusions about population by induction from samples); (ii) we have to have some more conditions to be satisfied to pool the data which will not be discussed here.

Hence the importance of this theorem is that the mean $\overline{X}$ of a random sample from any distribution is approximately normal with mean μ and variance σ^2 / n, if the sample size is large.

Due to this theorem, many variables can be considered following normal distribution as in many applications we are dealing with analysis involving the mean or the sum of many independent observations.

The practical aspect of the Central limit theorem is that it enables us to analyze and solve many problems which would be either impossible or difficult by other methods.

Applications of CLT

Example: Consider the examples presented in the last sections. The distribution of variable X is unknown and we have to decide the form of the distribution based on the sample frequency distribution. Consider the histogram of the examples. From their shape, we may want to fit a density function which is defined by this shape. Of course, there are an infinite number of density functions available for this shape. But usually we select the one which is of well-known forms in order to simplify the analysis. Therefore, we took a lognormal distribution as a suitable form.

Example: Consider a normal population distribution for which $\mu = 0$, $\sigma = 1$

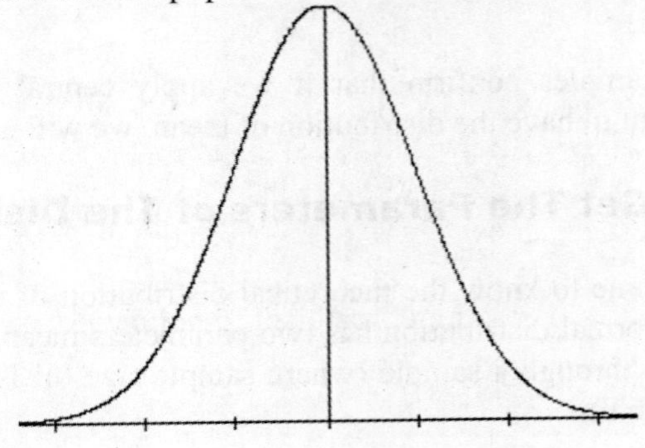

Figure: Normal distribution curve

A series of 100 random samples of 5 observations each was selected from this population.
The mean for each sample of 5 was calculated.
Then mean and sd of these 100 means was calculated as follows

      Mean of 100 means=0.015

      SD of 100 means = 0.452

Theoretically from CLT, mean should be 0 and sd of mean or SE should be $\sigma/\sqrt{n} = 1/\sqrt{5} = 0.4472$ which are very close to the above values.

Example: Let the distribution of the population is the uniform distribution with
Mean = 4.5
$\sigma = 2.87$
From this population, 100 random samples of 5 observations were selected and their means determined.
The mean of 100 means = 4.42
The SD of 100 means = 1.28

Theoretically from CLT, mean should be 4.5 and sd of mean or SE should be $\sigma/\sqrt{n} = 2.87/\sqrt{5} = 1.284$ which are very close to the experimental values.

Example: The age at onset for all cases of poliomyelitis in Massachusetts in 1949 has
Mean = 13.9 years
SD =10.3 years.
From this population, 100 random samples of 10 were selected.
The mean of 100 means = 13.8 years
The SD of 100 means= 3.5 years

If we apply CLT, mean should be 13.9 and sd = $\sigma/\sqrt{n}$ =3.42. Again we can see that whatever we get from actual experimental data, the same results we get from CLT.

All the above examples confirm that if we apply central limirt theorem or perform experiment to have the distribution of mean, we will get similar results.

How Do We Get The Parameters of The Distribution

Using CLT, we come to know the theoretical distribution of the data as normal distribution. But normal distribution has two parameters mean and variance. We approximate those through a sample (where sample size 'n' is sufficiently large or whatever available).

To summarize

- If we know the theoretical distribution, we can get approximate probabilities of different outcomes without performing the experiment
- Theoretical distributions are required as most of the statistical analyses proceed after the form of an appropriate theoretical distribution is decided upon
- Generally we select the one which is of well-known forms in order to simplify the analysis
- Due to Central limit theorem, the mean of a random sample from any distribution having mean μ and variance σ^2 is approximately normal with mean μ and variance σ^2/n, if the sample size is large
- The Central limit theorem enables us to analyze and solve many problems which would be either impossible or difficult by other methods

Do not put faith in what statistics say until you have carefully considered what they do not say.

— **William W. Watt**

Chapter 7

Statistical Inference: Basic Principles

Statistical inference may be viewed as the science of making conclusions about the population based on the information obtained from samples. Basically, inferential statistics is concerned with the methods that employ inductive reasoning i.e. on the basis of the sample we want to infer about the whole population.

Two main categories of problems in inference are
- Estimation
- Testing of hypothesis

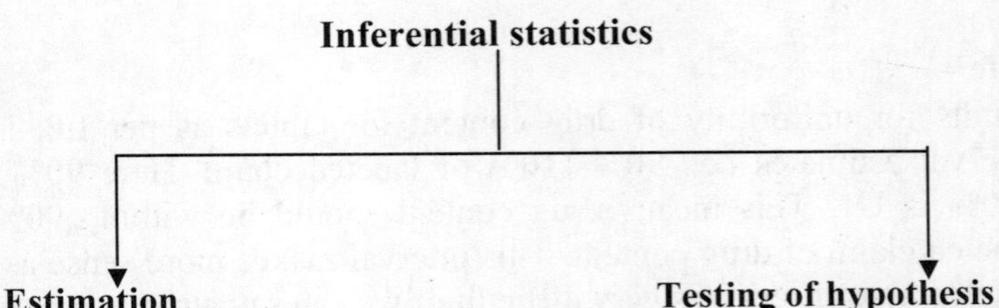

ESTIMATION

It has a meaning much like its meaning in ordinary usage. A population parameter is estimated based on a sample from that population. The estimation is of two types:
- Point Estimation
- Interval Estimation

Point Estimation

It estimates the unknown parameter as a single value based on the sample.
Example:

- Dissolution for isoniazid tablets I.P is 80% in 45 min. Here 80% is a point estimate of mean % dissolution in 45 min.
- Friability of some uncoated tablets is 1%. Here 1% is a point estimate of mean % friability.
- Disintegration time for sugar coated tablets as per I.P. is 60 min. This means tablets should be disintegrated totally in 60 min. Sixty (60) min is the point estimate of mean disintegration time.
- Disintegration time for soft-gelatin capsules as per I.P. is 60 min. This implies that capsules should be disintegrated totally in 60 min. Sixty (60) min is the point estimate of mean disintegration time.

Interval Estimation

In point estimation, estimate of the parameter is one value and hence it is very unlikely that the single point estimate will be precisely equal to the true value of the population parameter. Thus instead of point estimate, it is a good idea to have an interval which is likely to contain the required parameter, with some specified assurance. We calculate confidence interval (CI) for this purpose. This is known as interval estimation.

An interval estimate consists of two numerical values (L_1, U_1) defining an interval which includes the parameter being estimated with a specified degree of confidence i.e. probability. L_1 is the lower limit of the interval and U_1 is the upper limit. The confidence is given as $(1-\alpha)\%$.

Example

- Limits for uniformity of drug content for tablets as per I.P. is given as interval estimates i.e., 90%-110% of labeled claim. Here 90% is L1 and 110% is U1. This means drug content should be within 90%-110% of labeled claim of drug content. This interval makes more sense as meeting a single value criterion is very difficult or we can say almost impossible.
- Limits for uniformity of weight for tablets as per I.P. is given as interval estimates i.e., 85%-115% of mean of 20 tablets. In this case 85% is L_1 and 115% is U_1. This means tablet weight should be within 85%-115% of mean of 20 tablets.

Incorrect interpretation: A common mistake made by many people is to interpret confidence interval as there is $(1-\alpha)\%$ certainty that the parameter lies somewhere within these limits.

The correct interpretation of confidence interval is that there are $(1-\alpha)\%$ chances that the confidence interval will cover the true value of the

parameter. To understand this, assume an experiment is repeated a large number of times and a 95% confidence interval (CI) are computed each time. Then 95% confidence interval implies that 95% of these CIs will contain the true value of the population parameter.

Mathematically, it is written as

$$P\,(L \leq a \leq U) = 1-\alpha$$

Usually it is desirable for a confidence interval to have narrow interval and high confidence.

The width of the confidence interval decreases as
- The confidence level $(1-\alpha)$ decreases
- The sample size n increases
- The population standard deviation σ decreases

Confidence Interval for Parameter "Mean"

Let $x_1, x_2, x_3, \ldots\ldots, x_{n-1}, x_n$ is a random sample from a population having normal distribution with mean μ and variance σ^2 i.e. $N(\mu, \sigma^2)$. Then confidence interval for parameter "mean" with $(1-\alpha)\%$ confidence is given by as follows:

When standard deviation of the population is known:

$$(1-\alpha)\% \text{ confidence interval} = \bar{x} \pm \frac{Z_\alpha \sigma}{\sqrt{n}}$$

where $\bar{x}$ is the mean of the sample, σ is the population variance, n is the sample size and Z_α is a standard normal variate and is obtained from the normal distribution tables. Z_α can be one sided or two sided depending on whether interest is in only lower limit / upper limit or in both the limits.

When interest is in only lower limit or upper limit, one-sided Z_α should be taken i.e., we look for Z for probability "α" whereas when interest is in both the limits, two sided Z_α should be taken i.e., we look for probability "$\alpha/2$" in the table. Sometimes one sided and two sided Z_α are already defined in the table. In this case, we should look as per requirement.

When standard deviation of the population is unknown:

$$(1-\alpha)\% \text{ confidence interval} = \bar{x} \pm \frac{t_{\alpha,n-1}\,s}{\sqrt{n}}$$

where $\bar{x}$ is the mean of the sample, s is the estimate of population standard deviation (σ), n is the sample size and $t_{\alpha,n-1}$ is obtained from t distribution tables for α level of significance and n-1 degrees of freedom. $t_{\alpha,n-1}$ can be one sided or two sided depending on whether interest is in only lower limit / upper limit or in both the limits.

When interest is in only lower limit or upper limit, one-sided $t_{\alpha,n-1}$ should be taken i.e., we look for t for probability "α" whereas when interest is in both the limits, two sided $t_{\alpha,n-1}$ should be taken i.e., we look for probability "$\alpha/2$" in the table. Sometimes one sided and two sided $t_{\alpha,n-1}$ are already defined in the table. In this case, we should look as per requirement.

Example: Assay results for drug content of 10 randomly selected tablets are 101.8, 102.6, 99.8, 104.9, 103.8, 104.5, 100.7, 106.3, 100.6, and 105.0 mg. Calculate the 95% confidence interval of mean assay value (i) when sd is known and is 2.0 and (ii) when sd is unknown.

Solution: From the above sample data,
n = 10

Mean = $\bar{x}$ = 103.0
Standard deviation = 2.2

Our interest is in 95% confidence interval, that means α will be 1- 0.95 = 0.05. Also confidence interval should be two sided confidence interval as nothing is specified.

(i) When sd is known and is 2.0

$$95\% \text{ Confidence interval} = \bar{x} \pm \frac{Z_\alpha \sigma}{\sqrt{n}} = \bar{x} \pm \frac{Z_{0.05(2)} \sigma}{\sqrt{n}}$$

($Z_{0.05(2)}$ is 1.96 from the table)

$$= 103 \pm \frac{1.96 * 2.0}{\sqrt{10}}$$

$$= 103 \pm 1.2396$$

i.e. the 95% confidence interval of mean assay is (101.76, 104.24)

(ii) when sd is unknown, we have to use the standard deviation of the sample itself.

$$95\% \text{ Confidence interval} = \bar{x} \pm \frac{t_{\alpha,n-1} s}{\sqrt{n}} = \bar{x} \pm \frac{t_{0.05(2),9} s}{\sqrt{n}}$$

($t_{0.05(2),9}$ is 2.262 from the table)

$$= 103 \pm \frac{2.262 * 2.2}{\sqrt{10}}$$

$$= 103 \pm 1.5737$$

i.e. the 95% confidence interval of mean assay is (101.43, 104.57)

Example: Routine batches of isoniazid tablets which are made using tulsion 334 as disintegrant have mean disintegration time of tablet as 72 second. For cost effective reasons, it was proposed to replace tulsion with primojel but with the condition that there should not be significant change in disintegration time. So, a new batch with primojel was manufactured and a random sample of 10 tablets was tested for disintegration time. The data is as follows.

Tablet No.	Disintegration Time (Seconds)
1	59
2	72
3	58
4	65
5	77
6	83
7	72
8	77
9	62
10	62

Calculate point estimate (mean) and interval estimate (95% confidence interval) of mean disintegration time.

Solution:
From the data,
n = 10
Mean disintegration time of tablets = 68.7
Standard deviation = 8.67

Our interest is in 95% confidence interval, that means α will be 1- 0.95 = 0.05. Also confidence interval should be two sided confidence interval as nothing is specified.

As sd of population is not known,

$$95\% \text{ Confidence interval} = \bar{x} \pm \frac{t_{\alpha,n-1}s}{\sqrt{n}} = \bar{x} \pm \frac{t_{0.05(2),9}s}{\sqrt{n}}$$

($t_{0.05(2),9}$ is 2.262 from the table)

$$= 68.7 \pm \frac{2.262 * 8.67}{\sqrt{10}}$$

$$= 68.7 \pm 6.20$$

Hence point estimate (mean) of disintegration time of tablets is 68.7 seconds and 95% confidence limits for mean disintegration time are (62.5, 74.9). As this confidence interval covers mean disintegration time 72 seconds, tulsion can be replaced with primojel without significant change in disintegration time.

Confidence Interval for Proportion

$$(1-\alpha)\% \text{ confidence interval } = p \pm Z_\alpha \sqrt{\frac{pq}{n}}$$

where p is the proportion of the sample, $q = 1\text{-}p$, n = sample size and Z_α is a standard normal variate and is obtained from the normal distribution tables. Z_α can be one sided or two sided depending on whether interest is in only lower limit / upper limit or in both the limits.

When interest is in only lower limit or upper limit, one-sided Z_α should be taken i.e., we look for Z for probability "α" whereas when interest is in both the limits, two sided Z_α should be taken i.e., we look for probability "$\alpha/2$" in the table. Sometimes one sided and two sided Z_α are already defined in the table. In this case, we should look as per requirement.

Example: In a preclinical study, 100 untreated (control) animals were observed for the presence of liver disease. After 6 months, 25 of these animals were found to have the disease. Compute a 95% confidence interval for the true proportion of animals that would have this disease after 6 months if untreated.

Solution: From the above data
p = Proportion of animals having disease after 6 months

   = Number of animals found with disease after 6 months / Total number of animals observed

$$= \frac{25}{100} = 0.25$$

$q = 1\text{-}p = 1 - 0.25 = 0.75$

$n = 100$

Our interest is in 95% confidence interval, that means α will be $1 - 0.95 = 0.05$. Also confidence interval should be two sided confidence interval as nothing is specified.

$Z_{0.05(2)} = 1.96$ (From the normal distribution table)

95% Confidence interval for proportion of animals that would have this disease

$$\text{after 6 months if untreated} = p \pm Z_\alpha \sqrt{\frac{pq}{n}} = p \pm Z_{0.05(2)} \sqrt{\frac{pq}{n}}$$

$$= 0.25 \pm 1.96 \sqrt{\frac{0.25 * 0.75}{100}}$$

$$= 0.25 \pm 0.08487$$

Hence 95% confidence interval for the true proportion of animals that would have this disease after 6 months if untreated is (0.17, 0.33).

Example: Following table gives the frequency distribution of the volume of normal saline bottles from a sample of 100 normal saline bottles.

Volume range for normal saline bottles (ml)	Number of bottles (frequency)
527.52-532.5	05
532.5-537.5	18
537.5-542.5	42
542.5-547.5	27
547.5-552.5	08
Total	100

From this data, calculate point estimate "mean" and interval estimate "99% confidence interval" of volume of normal saline bottles.

Solution:

As volume of bottles are given as frequency table, average volume of bottles will be calculated using formulae for frequency distribution.

Weighted mean = 540.75

Standard deviation = s = 4.87

n = 100

Our interest is in 99% confidence interval, that means α will be 1- 0.99 = 0.01. Also confidence interval should be two sided confidence interval as nothing is specified.

As standard deviation of population is not known,

$$99\% \text{ Confidence interval} = \bar{x} \pm \frac{t_{\alpha,n-1}s}{\sqrt{n}} = \bar{x} \pm \frac{s}{\sqrt{n}} t_{0.01(2),99}$$

$$(\ t_{0.01(2),99} \text{ is } 2.627 \text{ from the table})$$

$$= 540.75 \pm \frac{4.87}{\sqrt{100}} \times 2.627$$

$$= 540.75 \pm 1.2793$$

Hence the point estimate of volume of normal saline bottles is 540.75 ml and an interval estimate of the volume of normal saline bottles is (539.47-542.03) with confidence 99%.

Example: An improved method for somatic plantlet production in hybrid larch (larix) is developed. Using this method, an experiment was conducted with 25 plants. Numbers of mature somatic embryos were obtained in the 25 plants as follows:

58	137	69	61	42
66	83	38	29	25
188	32	83	162	102
19	51	130	77	153
47	74	43	49	221

Find out 95% confidence interval for the average number of mature somatic embryos.

Solution:
From the data,
n = 25
Mean number of mature somatic embryos = 83.96 $\cong$ 84 (as number of cells has to be whole number),
standard deviation (sd) = 52.46,

Our interest is in 95% confidence interval, that means α will be 1- 0.95 = 0.05. Also confidence interval should be two sided confidence interval as nothing is specified.

$$95\% \text{ Confidence interval} = \bar{x} \pm \frac{t_{\alpha,n-1} s}{\sqrt{n}} = \bar{x} \pm \frac{s}{\sqrt{n}} t_{0.05(2),24}$$

($t_{0.05(2),24}$ is 2.064 from the table)

$$= 83.96 \pm \frac{2.064 * 52.46}{\sqrt{25}}$$

$$= 83.96 \pm 21.6554$$

i.e. the 95% confidence interval of the average number of mature somatic embryos is (62, 106)

All generalizations are false, including this one. Statisticians do it with 95% confidence.   -Anonymous

BASIC IDEAS OF HYPOTHESIS TESTING

More often in practice, we are not only concerned about the problem of estimating, but also with deciding whether an unknown parameter of a population is equal to a certain specified value or an unknown parameter of a population is same as of another population. The procedure which leads to decide upon these, i.e., accepting or rejecting specified statements about the population is called the testing of hypothesis. Such decisions are called statistical decision.

For example, we may wish to decide on the basis of sample data whether
- the weight of the tablets of a batch meets the target weight
- the design of the tablet is optimum as per the requirement
- a new drug is really effective in curing a disease
- a new drug is better than the existing drug for a particular illness

Testing of hypothesis is one of the most important concepts in statistics. It is known as "testing of hypothesis" because we set a hypothesis about the population and test whether that hypothesis is true or not by statistical tests based on the sample. This hypothesis is nothing but an assumption about the population under study. For example,
- A hypothesis for the average weight of tablets of a batch may be "the average weight of the tablet is 325 mg".
- A hypothesis for the cure rate of a drug may be "the cure rate of the drug is 0.75".

Since the population is specified by one or more parameters of its distribution (for example, Binomial distribution is defined by n and p (mean "np" and variance "npq"), Poisson distribution is defined by λ (mean and variance both as λ), normal distribution is defined by mean "μ" and variance "σ^2"), a hypothesis is an assumption regarding parameter of a population distribution. In the above examples, parameters are "mean" and "rate".

Testing of hypothesis comprises of:
- The hypothesis being tested is often represented by H_0 called the null hypothesis
- The alternative hypothesis denoted by H_1 which is accepted in case H_0 is rejected
- A decision rule called as a **statistical test** which leads to acceptance or rejection of a hypothesis

A **statistical test** is based on
- information obtained from sample in the form of a **test statistic** which is a single valued function of the observations
- a criterion for rejecting H_0 based on the test statistic and the amount of evidence required to accept that the event is unlikely. This "amount of evidence" is calculated based on the distribution of 'test statistic' (As different samples from the same population will give different values of test statistic, the test statistic have their own distributions). This 'amount of evidence' is basically an error that can be tolerated in making decision of rejection of a hypothesis. This error is known as "level of significance" or "significance level".

"Level of significance" or "Significance level" is a number that expresses the probability that the results of a given experiment or study could have occurred purely by chance not because that the hypothesis is true. This number can be considered as margin of error.

As per the chosen significance level, we can determine a region such that if the test statistic falls in the region, we reject H_0. This region of the distribution is called the rejection region or critical region. The acceptance region for H_0 is given by the complement of the critical region.

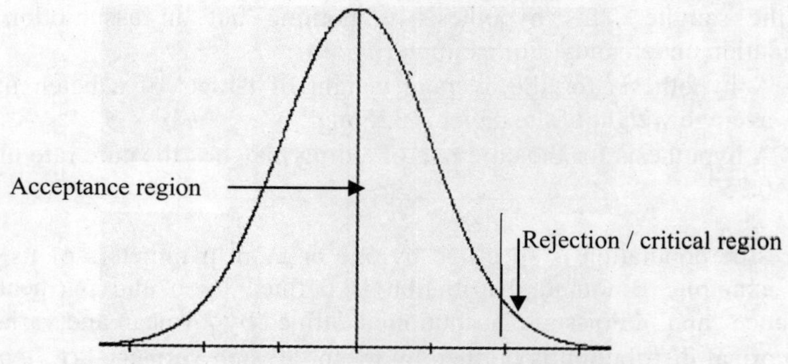

Figure: Figure illustrating the concept of "P" value

Operationally, if the probability of happening the event of null hypothesis is less than the significance level, it is probably occurred by chance and a result is called statistically significant. This probability of happening the event of null hypothesis is known as **P value**

P value
- Whether hypothesis is rejected or accepted, always give "P" values it gives an idea of how strongly the data contradict (or support) the hypothesis.
- Knowing "P" value allows us to assess significance at any level.
- Smaller the "P" value is, stronger is the evidence against H_0 provided by the data.

Always keep in mind that by acceptance of a hypothesis, we do not mean that it is proved to be true. All that implied is that in the light of the given sample data, we find no reason to question the validity of the hypothesis.

Various phases of "Testing of hypothesis"
- Definition of the problem
- Method selection
- The interpretation of the results

Definition of the Problem

- This is very important as based on this, further steps are taken.
- There should be clearly defined objectives / questions that the experiment is designed to answer, for example, objective can be to determine the efficacy and safety of X Product in the prevention of Y disease.
- We should be clear in our mind what we want. For example if two treatments have to be compared, it should be very clear whether interest is in equality of two treatments or superiority of one treatment over the other one.

In first case we want to test the hypothesis:

$$H_0 : \mu_1 = \mu_2 \quad \text{vs.} \quad H_1 : \mu_1 \neq \mu_2$$

whereas in the second case we are interested in

$$H_0 : \mu_1 = \mu_2 \quad \text{vs.} \quad H_1 : \mu_1 > \mu_2$$

where μ_1 and μ_2 are the efficacy parameters of the two treatments. Later on we will see how the test results can change based on the definition of the problem, i.e. in first case, H_0 can be rejected and in second case it can be accepted based on the same sample data.

Method Selection

It consists of two steps
- Design of experiment
- Analysis of the data

Design of Experiment
Designing an experiment means deciding
- What measurements should be taken
- How the measurements should be taken
to answer a particular question in a valid, efficient and economical way.

When experiments are designed with the objectives in mind, researchers can identify those sources that they consider can influence the outcome and choose a design that will allow them to measure the extent of the contribution of those sources to the outcomes / measurements.

Hence the questions which should be answered before performing the experiment are
- What the appropriate procedure (design) should be for conducting an experiment.
- In which manner the data should be collected for the specific purpose i.e. designs should be decided in advance.

- What data has to be measured for the purpose.

"Failing to Plan is Planning to Fail"

Example: If we have to perform a clinical trial to compare test and reference drugs, following sources can be taken care of, if it is known that these factors play a role in giving the effect:
- Sex
- Age
- Disease condition

Test and reference drugs should be allocated to subjects in such a way that the subjects' groups are homogenous in all the above aspects for test as well as reference groups.

Example: If we have to perform an experiment in animals to compare the toxicity of test and reference drugs, following sources of variation should be taken care of as these all may influence the outcome of the experiment:
- Stain
- Sex
- Body weight,
- Age
- Housing and environment conditions

Test and reference groups of animals should be homogenous in all the above aspects so that the above factors influence the toxicity outcomes in the similar way in both the drugs' group.

Some Fundamental Points of Study Design

Control: One should see if any yardstick is required in the experiment to properly interpret the results. Commonly used word for this is "control" in Pharmaceutical Sciences / drug discovery and development. A control or control group is a group of units / subjects which is other than the group of units / subjects assigned to the test treatment. The control is used against which the test can be compared.

If we are interested in, for example,
- Knowing the average weight or assay of tablet,
- Proportion of tablets defective in a batch,
- Performing a dose escalation study,

No need to have controls.

But if we are concerned with, for example,
- Looking at the dissolution profile of a new formulation of a drug for which another formulation is already in the market
- Looking at the pharmacological effects of a drug in animals
- Testing the efficacy of a formulation or new drug

We need to have controls.

Controls are of the following types:
- Simple control
- Placebo control
- Active control

Simple Control: The control group receives no treatment.

Placebo Control: The control group receives a treatment that is identical in outward appearance to the test treatment, but has no drug. Examples: sugar pills, saline injections, etc.

Active Control: The control group receives a different treatment (often the current standard treatment) for the same therapeutic area as of the test drug.

Randomization: Randomization means treatments should be allocated to the subjects / experimental units randomly i.e. each subject / experimental unit should have equal probability of getting any of the investigational treatments. This is the preferred way of assigning subjects / experimental units to control and treatment groups.

Advantages:
- *Eliminates bias*
- *Ensures comparable study groups for test and control treatments*
- *Allows for use of valid statistical tests*

Disadvantages:
- Requires in advance the number of subjects / experimental units in each treatment group for treatment allocation (usually equal allocation for all the treatments as power of the statistical tests are maximum for this case)

Replication: It implies that each treatment should be applied to more than one subjects / experimental units within the experiment.

Advantages:
- Reduces chance variation
- Measures variability
- Allows for use of valid statistical tests

Blinding: Blinding is the process of masking the treatment assignment to study investigators and / or experimental units (for example, participants). This process helps to control the potential for bias because individuals tend to change their behaviors based on treatment information. For example, if somebody is given the test product and he / she knows this, there are very high chances that he / she will give response of less efficacy and more side effects with the test product as it is human tendency to be skeptical about new things.

Blinding is of the following types:

Unblinded / Open Label: Patient and investigator, both are aware about the treatment allocated.
Advantage:
- Simpler to execute
- Better reflects actual clinical practice

Disadvantages:
- May introduces bias in outcomes

Single Blind: Patient is blinded to treatment but investigator is not.
Advantage:
- Puts investigators at ease
- Easier to administer

Disadvantages:
- May introduce investigator bias in outcomes

Double Blind: Patient and investigator, both are blinded to treatment.
Advantage:
- Reduces risk of bias

Disadvantages:
- Requires outside personnel to monitor safety and treatment allocation
- May be very difficult to accomplish in studies of some treatments

Types of Designs

There are various designs available in statistical literature. For example:
Parallel group design
Randomized block designs
Nonrandomized concurrent / historical controls designs
Cross-over designs
Factorial designs
Sequential designs

Any of these designs can be applied based on the objectives of the study. To get details on designs, many books on design aspects are available (For example, Biostatistical analysis by J H Zar, Biopharmaceutical Statistics by Bolton).

What Has to be Measured

Outcomes / Response variable: Measurements taken during the course of the trial are the outcomes of the experiment. Examples,
- A lab measurement
- Total mortality
- Symptomatic relief
- Death from a specific cause
- Incidence of a disease
- A specific adverse effect

Endpoints: Based on the questions, endpoints are defined based on the outcomes of the experiment in order to accomplish the objectives of a study.
- – In general, a single endpoint identified for the primary question
- – Endpoint must be capable of being assessed in all participants, and should be measured the same way.
- – Endpoint should be capable of unbiased assessment
- – Examples
    - – Observe friability of formulation to see the effect of humidity
    - – Percentage of patients with a reduction in pain
    - – Proportion of patients surviving two years post-treatment
    - – Average length of survival of patients post-treatment
    - – Change in body / organ weight before and after the treatment is given

Analysis of the data

Different types of statistical analysis may be appropriate depending on design (for example parallel group, cross-over design, etc.) and the sort of outcome measure (for example, mean, proportion, etc.). For a particular design and type of outcome measure, there is a definite way of analyzing the data i.e. the design and the analysis go together.

Hence one should be careful in choosing the appropriate statistical test based on design and outcome.

Interpretation of the results

Always remember that statistical significance is not clinical significance. Hence interpretation of the results is a very important step in testing of hypothesis.

Misinterpretation is the most deadly of human sins.

-Lester del Rey

TWO TYPES OF ERROR

Whenever we perform any statistical test on the data, there are chances of making error i.e., there is a chance that null hypothesis is true but we reject the null hypothesis or that null hypothesis is false but we accept the null hypothesis. This is summarized in the following table:

		Decision based on the statistical test	
		Accept H_0	**Reject H_0**
True situation	**H_0 Correct**	Correct	Incorrect (Type I Error (α))
	H_0 false (H_1 Correct)	Incorrect (Type II Error (β))	Correct

When H_0 is correct and we accept H_0 based on the statistical test or when H_0 is not correct and we reject H_0 based on the statistical test, it is correct and hence no problem.

But when H_0 is correct and we reject H_0 based on the statistical test or when H_0 is not correct and we accept H_0 based on the statistical test, errors occur. These two errors are defined as follows:

Type I Error   = Probability (Reject H_0 / H_0 is true)
                   = Level of significance
                   = α

Type II Error   = Probability (Accept H_0 / H_0 is false)
                   = β

- Ideally we would like to choose a test that minimizes both the errors. But this is not possible, if we minimize type I error, type II error increases and vice versa.
- So normally we fix type I error (α) at a level and try to minimize type II error (β). The choice of level of significance depends on the experiment itself. If rejection of the null hypothesis when it is true is a serious error, level of significance should be a small value. If it is other way, taking a value of level of significance as high as 0.1 is also fine.
- In general, we use type one error or level of significance as 5%.

POWER

Whenever we perform any statistical test on any data, our interest is in detecting the difference, if it exists between investigational treatments. Power is the probability of detecting a difference when there is actually a difference.

Power = Probability (declare the investigational treatments are different when those are actually different)
         = 1 – P(declare the investigational treatments are same when those are different)
         = 1- Probability (Accept H_0 / H_0 is false)
         = 1 – Type II Error

Let us summarize testing of hypothesis as follows:

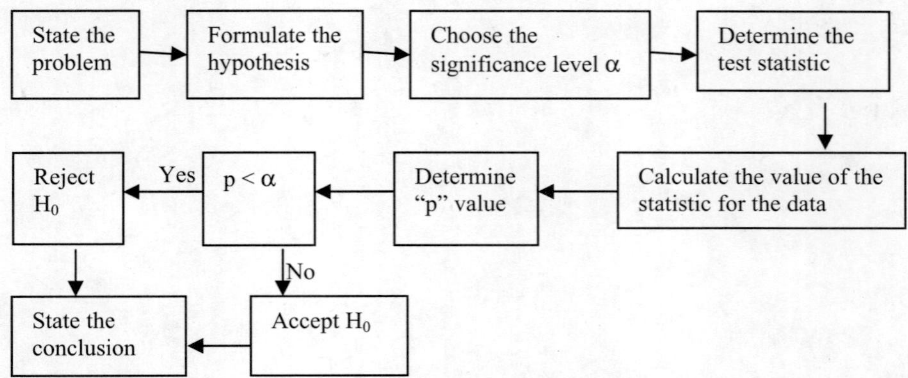

Figure: Typical steps in statistical test

In the next chapters, we will discuss some tests for populations having normal distribution. The reasons for choosing only normal distribution are

- The methods based on normal distribution are extensive and well developed to facilitate and simplify the solutions of many practical problems.
- This distribution is suitable, at least approximately, for a large number of practical problems in pharmaceutical / medical sciences.

To summarize

- Estimation is of two types: point and interval
- Testing of hypothesis has three important aspects
    o Definition of the problem
    o Method selection
        ▪ Designing an experiment
        ▪ Method of analysis
    o Interpretation of the results
- The fundamental points in Designing an experiment
    o Controls
    o Randomization
    o Replication
    o Blinding

No one can guarantee that the results of any piece of research will prove useful, but at least it should not be doomed from the start to prove nothing

— Anonymous

Chapter 8

Basic Statistical Tests for Population Having Normal Distribution

ONE SAMPLE HYPOTHESIS

Let $x_1, x_2, x_3, \ldots\ldots, x_{n-1}, x_n$ is a random sample from a population having normal distribution with mean μ and variance σ^2 i.e. $N(\mu, \sigma^2)$.

Our interest is to test whether the mean μ is equal to a value μ_0. To understand the concept let us consider an example of tablet manufacturing. When tablets are manufactured, there is always a target weight of the tablet. So our interest will be in testing whether the mean weight of manufactured tablets is equal to the target weight or not.

Tests for mean when σ^2 is known

Let population variance (σ^2) is known. There are various tests known as Z tests available in literature for different–2 situations, depending upon the objective of the experiment.

(i) If the null and alternative hypotheses are
$$H_0 : \mu = \mu_0 \qquad \text{vs.} \qquad H_1 : \mu \neq \mu_0$$

Our interest is to know whether the mean is equal to μ_0 versus mean is not equal to μ_0 i.e. mean can be less than μ_0 or mean can be more than μ_0. That means we are testing on both sides, hence tests with this type of alternative hypothesis are known as two tailed tests or two sided tests.

The null hypothesis will be rejected at α level of significance if

$$Z = \frac{\bar{x} - \mu_0}{\sigma/\sqrt{n}} > z_{\alpha/2} \qquad \text{or} \qquad Z = \frac{\bar{x} - \mu_0}{\sigma/\sqrt{n}} < -z_{\alpha/2}$$

where $\bar{x}$ is the sample mean, n is the sample size, $Z_{\alpha/2}$ is a standard normal variate and is obtained from the normal distribution tables. We have to look into the table for two sided Z with α level of significance. It is also represented by $Z_{\alpha(2)}$.

The function "Z" of sample and population parameters is known as test statistic.

(ii) If the null and alternative hypotheses are

$$H_0 : \mu \geq \mu_0 \qquad \text{vs.} \qquad H_1 : \mu < \mu_0$$

Our interest is to know whether the mean is equal to or greater than μ_0 versus mean is less than μ_0 i.e. we are testing only on one side (lower side), hence tests with this type of alternative hypothesis are known as single tail tests or single sided tests.

The null hypothesis will be rejected at α level of significance if

$$Z = \frac{\bar{x} - \mu_0}{\sigma/\sqrt{n}} < -z_{\alpha}$$

where $\bar{x}$ is the sample mean, n is the sample size, $-z_{\alpha}$ is a standard normal variate and is obtained from the normal distribution tables. We have to look into the table for one sided Z with α level of significance. It is also represented by $-Z_{\alpha(1)}$.

(iii) If the null and alternative hypotheses are

$$H_0 : \mu \leq \mu_0 \qquad \text{vs.} \qquad H_1 : \mu > \mu_0$$

Our interest is to know whether the mean is equal to or less than μ_0 versus mean is more than μ_0 i.e. we are testing only on one side (upper side), hence tests with this type of alternative hypothesis are known as single tail tests or single sided tests.

The null hypothesis will be rejected at α level of significance if

$$Z = \frac{\overline{x} - \mu_0}{\sigma/\sqrt{n}} > z_\alpha$$

where $\overline{x}$ is the sample mean, n is the sample size, z_α is a standard normal variate and is obtained from the normal distribution tables. We have to look into the table for one sided Z with α level of significance. It is also represented by $z_{\alpha(1)}$.

Tests for mean when σ^2 is unknown

Let population variance (σ^2) is not known and sample variance (s^2) is an estimate of σ^2. There are various tests known as t tests available in literature for different–2 situations, depending upon the objective of the experiment.

(i) If the null and alternative hypotheses are

$$H_0 : \mu = \mu_0 \qquad \text{vs.} \qquad H_0 : \mu \neq \mu_0$$

The null hypothesis will be rejected at α level of significance if

$$t = \frac{\overline{x} - \mu_0}{s/\sqrt{n}} > t_{\alpha/2,n-1} \qquad \text{or} \qquad t = \frac{\overline{x} - \mu_0}{s/\sqrt{n}} < -t_{\alpha/2,n-1}$$

where $\overline{x}$ is the sample mean, n is the sample size, $t_{\alpha/2,n-1}$ is a "t variate" and is obtained from the "t distribution tables". It is also represented by $t_{\alpha(2),n-1}$. We have to look into the table for two sided t with α level of significance and n-1 degrees of freedom.

(ii) If the null and alternative hypotheses are

$$H_0 : \mu \geq \mu_0 \qquad \text{vs.} \qquad H_0 : \mu < \mu_0$$

The null hypothesis will be rejected at α level of significance if

$$t = \frac{\overline{x} - \mu_0}{s/\sqrt{n}} < -t_{\alpha,n-1}$$

where $\overline{x}$ is the sample mean, n is the sample size, $t_{\alpha,n-1}$ is a t variate and is obtained from the "t distribution" tables. We have to look into the table for one sided t with α level of significance and n-1 degrees of freedom. It is also represented by $t_{\alpha(1),n-1}$.

(iii) If the null and alternative hypotheses are

$$H_0 : \mu \leq \mu_0 \qquad\qquad vs. \qquad H_0 : \mu > \mu_0$$

The null hypothesis will be rejected at α level of significance if

$$t = \frac{\overline{x} - \mu_0}{s/\sqrt{n}} > t_{\alpha, n-1}$$

where $\overline{x}$ is the sample mean, n is the sample size, $t_{\alpha, n-1}$ is a t variate and is obtained from the "t distribution" tables. We have to look into the table for one sided t with α level of significance and n-1 degrees of freedom. It is also represented by $t_{\alpha(1), n-1}$.

Example: Three exhibit batches of tablet X of 5 mg, of size 10,0000 each, have been already taken and it has been established that the average drug content/ assay for tablet X is 5.01 mg and its standard deviation is 0.11. Regular batch production of the same was initiated after approval from regulatory authorities. Afterwards, the first batch with the same manufacturing process was manufactured. Twenty tablets from this batch were analyzed for the uniformity of drug content (as per pharmacopoeia) and the results are as follows:
5.13, 4.98, 5.20, 5.08, 4.99, 5.04, 5.03, 5.08, 5.06, 5.17, 5.09, 5.01, 4.96, 5.02, 5.06, 5.00, 4.99, 5.18, 5.24, and 5.00
The objective is to determine whether the average drug content of the new batch is same as that of exhibit batches.

Solution:
For the problem in hand, data from exhibit batches is taken as population data. Hence mean and sd for the population are
Mean $= \mu_0 = 5.01$ and sd $= \sigma = 0.11$

Let μ be the average drug content of tablets population from which the sample was taken. Hence the null hypothesis and alternative hypothesis for the above problem will be:

$$H_0 : \mu = 5.01 \qquad\qquad vs. \qquad H_0 : \mu \neq 5.01$$

As standard deviation of the population is known and our interest is in whether the drug content is less than or more than 5.01, two sided Z test will be applied.

From the new batch
n = 20, Mean $= \overline{x} = 5.0655$

On applying one sample two-tailed Z-test,

$$Z = \frac{\bar{x} - \mu_0}{\sigma/\sqrt{n}} = \frac{5.0655 - 5.01}{0.11/\sqrt{20}} = 2.256$$

The Z value for a two-sided test at the 5 % level is 1.96 from the standard normal distribution table.

As the calculated Z value is greater than tabulated Z value, null hypothesis is rejected and alternative hypothesis is accepted. It infers that the new batch produced shows significant deviation in the average drug content, from the exhibit batches. Also P=0.0241 which is <0.05.

Example: Routine batches of isoniazid tablets which are made using tulsion 334 as disintegrant have mean disintegration time as 72 second. For cost effective reasons, it was decided to replace tulsion with primojel. So, a new batch with primojel was manufactured and a random sample of 10 tablets was tested to find disintegration time. The data is as follows

Tablet No.	Disintegration Time (Sec)
1	59
2	72
3	58
4	65
5	77
6	83
7	72
8	77
9	62
10	62

Test whether replacing tulsion with primojel is without significant change in disintegration time of the tablet.

Solution:
Our interest is to test whether mean disintegration time of tablets is 72 seconds or not. Here the parameter of interest is disintegration time of tablets, i.e., $\mu=$ Mean disintegration time of tablets and $\mu_0= 72$ seconds. As our interest is to know if mean disintegration time is less than or greater than 72 sec, null and alternative hypotheses are

$$H_0 : \mu = 72 \qquad \text{vs.} \qquad H_1 : \mu \neq 72$$

Let mean and standard deviation of disintegration time of tablets are $\bar{x}$ and s. From the above data

$\bar{x} = 68.7$

$s = 8.67$

$n = 10$

Also the population variance is not known. Hence we will perform 2 tailed t test. The test is performed at 5% level of significance.

$$t = \frac{\bar{x} - \mu_0}{s/\sqrt{n}} = \frac{68.7 - 72}{8.67/\sqrt{10}} = \frac{-3.3}{8.67/3.16} = \frac{-10.44}{8.67} = -1.20$$

From "t" table, for n-1=9 degrees of freedom and at 5% level of significance, $t_{\alpha/2,n-1} = 2.262$ and hence $-t_{\alpha/2,n-1} = -2.262$.

As t calculated (-1.20) is more than $-t_{\alpha/2}$ (= -2.262), the null hypothesis will be accepted at 5% level of significance. Also P = 0.261 which is >0.05.

Hence mean disintegration time of isoniazid tablets with primojel is comparable with mean disintegration time of tablets with tulsion based on the given sample. So tulsion can be replaced with primojel without significant change in disintegration time of the tablet.

Example: A pharmaceutical company is concerned about filling its bottles with either too little or too much medicine. A filling machine which has been set to fill 8-ounce bottles is checked by selecting at random 9 bottles and measuring their contents. The results are: Mean = 8.05 and sd= 0.035.
Assuming that the quantity filled by the machine is normally distributed, test the hypothesis that the mean is 8 ounces.

Solution:
Our interest is to test whether mean quantity filled by the machine is 8 ounce or not. Here the parameter of interest is mean quantity filled by the machine, i.e., μ= mean quantity filled by the machine and μ_0= 8 ounce. As our interest is to know if mean quantity filled by the machine is less than or greater than 8 ounce, null and alternative hypotheses are

$H_0 : \mu = 8$ vs. $H_1 : \mu \neq 8$

Let mean and standard deviation of quantity filled by the machine are $\bar{x}$ and s. From the above data

$\bar{x} = 8.05$

$s = 0.035$

$n = 9$

Also the population variance is not known. Hence we will perform 2 tailed t test. The test is performed at 5% level of significance.

$$t = \frac{\overline{x} - \mu_0}{s/\sqrt{n}} = \frac{8.05 - 8}{0.035/\sqrt{9}} = \frac{0.05}{0.035/3} = \frac{0.15}{0.035} = 4.29$$

From "t" table for n-1=8 degrees of freedom and at 5% level of significance, $t_{\alpha(2),n-1} = 2.306$.

As t calculated (=4.29) is more than $t_{\alpha(2),n-1}$ (= 2.306), the null hypothesis will be rejected at 5% level of significance. Also P < 0.01 which is < 0.05.

Hence mean quantity filled by the machine is statistically significantly different from 8 ounces based on the given sample.

Example: The mean systolic blood pressure for white males of 35 to 44 years age is 127 mm hg and the standard deviations in this population are 7. The medical director of a company looks at the medical records of 72 company executives in this age group and finds that the mean systolic blood pressure in this sample is 126.1 mm hg. Is this evidence that executives' blood pressure differ from the national average?

Solution.
The mean systolic blood pressure of population = 127 mm
Standard deviation of population = 7

Here the parameter of interest is mean systolic blood pressure, i.e., μ= Mean systolic blood pressure and μ_0= 127 mm. As our interest is to know if mean systolic blood pressure is less than or greater than 127 mm, null and alternative hypotheses are
$$H_0 : \mu = 127 \qquad \text{vs.} \qquad H_1 : \mu \neq 127$$

Let mean of mean systolic blood pressure is $\overline{x}$. From the above data
$\overline{x} = 126.1$
$n = 9$

As standard deviation of the population is known and our interest is to test whether executives' blood pressures differ from the national average, two sided Z test will be applied. The test is performed at 5% level of significance.

$$Z = \frac{\overline{x} - \mu_0}{\sigma/\sqrt{n}} = \frac{126.1 - 127}{7/\sqrt{72}} = 1.09$$

The Z value for a two-sided test at the 5 % level of significance is 1.96 from the standard normal distribution table. As the calculated Z (=1.09) value is less than tabulated Z value (=1.96), null hypothesis is accepted and alternative hypothesis is rejected. It infers that there is no evidence that executives' blood pressures differ from the national average based on this sample (P=0.2758 which is >0.05).

TWO SAMPLE HYPOTHESIS

Let there are two random samples as follows:

	Sample 1	Sample 2
	$(x_{11}, x_{12}, ..., x_{1n_1})$	$(x_{21}, x_{22}, ..., x_{2n_2})$
Mean	$\overline{x}_1$	$\overline{x}_2$
Sample Variance	s_1^2	s_2^2
No. of data Points	n_1	n_2

Also assume that these two samples are drawn from population having variances σ_1^2 and σ_2^2.

Our objective is to test whether these two samples are from the population with same mean i.e., whether both the means are same.

If following assumptions are met,
1. Both the samples are from normal distributions
2. Observations are independent,

One of the following tests can be applied:

Tests for means when σ^2 is known

(i) If the null and alternative hypotheses are

$$H_0 : \mu_1 = \mu_2 \qquad \text{vs.} \qquad H_1 : \mu_1 \neq \mu_2$$

The null hypothesis will be rejected at α level of significance if

$$Z = \frac{\overline{x}_1 - \overline{x}_2}{\sqrt{\dfrac{\sigma_1^2}{n_1} + \dfrac{\sigma_2^2}{n_2}}} > z_{\alpha/2} \qquad \text{or} \qquad Z = \frac{\overline{x}_1 - \overline{x}_2}{\sqrt{\dfrac{\sigma_1^2}{n_1} + \dfrac{\sigma_2^2}{n_2}}} < -z_{\alpha/2}$$

$Z_{\alpha/2}$ is same as defined in one sample case.

(ii) If the null and alternative hypotheses are

$$H_0 : \mu_1 \geq \mu_2 \qquad \text{vs.} \qquad H_1 : \mu_1 < \mu_2$$

The null hypothesis will be rejected at α level of significance if

$$Z = \frac{\overline{x}_1 - \overline{x}_2}{\sqrt{\dfrac{\sigma_1^2}{n_1} + \dfrac{\sigma_2^2}{n_2}}} < -z_\alpha$$

Z_α is same as defined in one sample case.

(iii) If the null and alternative hypotheses are

$$H_0 : \mu_1 \leq \mu_2 \qquad \text{vs.} \qquad H_1 : \mu_1 > \mu_2$$

The null hypothesis will be rejected at α level of significance if

$$Z = \frac{\overline{x}_1 - \overline{x}_2}{\sqrt{\dfrac{\sigma_1^2}{n_1} + \dfrac{\sigma_2^2}{n_2}}} > z_\alpha$$

Z_α is same as defined in one sample case.

Tests for means when σ^2 is unknown

As population variance is unknown, it is estimated by sample variance

$$s_i^2 = \frac{ss_i}{(n_i - 1)} \quad \text{............................for } i = 1 \text{ and } 2.$$

where $ss_i = \sum_{j=1}^{n_i} x_{ij}^2 - n_i * \overline{x}_i$,

s_1^2 is the variance of sample 1 and s_2^2 is the variance of sample 2.

For this case, first we have to test whether both the samples are from the populations with same variance and statistical test for mean will be chosen based on the outcome of this test.

Comparison of variances of two samples from populations having normal distribution: F test
Hypothesis for the equality of variances will be

$$H_0 : \sigma_1^2 = \sigma_2^2 \qquad \text{vs.} \qquad H_1 : \sigma_1^2 \neq \sigma_2^2$$

We calculate F statistic as follows:

$$F = s_1^2 / s_2^2$$

This F statistic has F distribution with $(n_1 - 1, n_2 - 1)$ df.

So we compare this calculated F statistic ($= F_{\text{calculated}} = F_{\text{cal}}$) with the tabulated value of F ($= F_{\text{tabulated}} = F_{\text{tab}}$) with $(n_1 - 1, n_2 - 1)$ df at α level of significance (two sided). Reject H_0 if

$$F_{\text{calculated}} > F_{\text{tabulated}} \;(= F_{n_1-1, n_2-1, \alpha(2)}) \text{ or } F_{\text{calculated}} < F_{\text{tabulated}} \;(= F_{n_1-1, n_2-1, (1-\alpha(2))})$$

Tables available in most of the books give $F_{n_1-1, n_2-1, \alpha(2)}$. However, $F_{n_1-1, n_2-1, (1-\alpha(2))}$ can be calculated as follows:

$$F_{n_1-1, n_2-1, (1-\alpha(2))} = \frac{1}{F_{n_1-1, n_2-1, \alpha(2)}}$$

Otherwise a simple approach is to keep the larger sample variance in the numerator and if calculated F is greater than the tabulated F, reject the null hypothesis of equality of variances otherwise accept.

Based on the F test outcome, appropriate test will be applied for testing means as follows:

- If null hypothesis of equality of variances is accepted, two sample t test should be applied.
- If null hypothesis of equality of variances is rejected, Aspin - Walch test should be applied.

Two sample t test for means when $\sigma_1 = \sigma_2$

Before applying the 2 sample t test, calculate the pooled variance of the samples as follows:

$$s_p^2 = \frac{(n_1 - 1)s_1^2 + (n_2 - 1)s_2^2}{(n_1 - 1) + (n_2 - 1)}$$

(i) If the null and alternative hypotheses are

$$H_0 : \mu_1 = \mu_2 \qquad \text{vs.} \qquad H_1 : \mu_1 \neq \mu_2$$

The null hypothesis will be rejected at α level of significance if

$$t = \frac{\overline{x}_1 - \overline{x}_2}{s_p \sqrt{\dfrac{1}{n_1} + \dfrac{1}{n_2}}} > t_{\alpha/2,\,n_1+n_2-2} \qquad \text{or} \qquad t = \frac{\overline{x}_1 - \overline{x}_2}{s_p \sqrt{\dfrac{1}{n_1} + \dfrac{1}{n_2}}} < -t_{\alpha/2,\,n_1+n_2-2}$$

where $t_{\alpha/2,\,n_1+n_2-2}$ is a two sided t variate with α level of significance and $n_1 + n_2 - 2$ degrees of freedom and is obtained from "t distribution" tables.

(ii) If the null and alternative hypotheses are

$$H_0 : \mu_1 \geq \mu_2 \qquad\qquad \text{vs.} \qquad H_1 : \mu_1 < \mu_2$$

The null hypothesis will be rejected at α level of significance if

$$t = \frac{\overline{x}_1 - \overline{x}_2}{s_p \sqrt{\dfrac{1}{n_1} + \dfrac{1}{n_2}}} < -t_{\alpha,\,n_1+n_2-2}$$

where $t_{\alpha,\,n_1+n_2-2}$ is a one sided t variate with α level of significance and $n_1 + n_2 - 2$ degrees of freedom and is obtained from "t distribution" tables.

(iii) If the null and alternative hypotheses are

$$H_0 : \mu_1 \leq \mu_2 \qquad\qquad \text{vs.} \qquad H_1 : \mu_1 > \mu_2$$

The null hypothesis will be rejected at α level of significance if

$$t = \frac{\overline{x}_1 - \overline{x}_2}{s_p \sqrt{\dfrac{1}{n_1} + \dfrac{1}{n_2}}} > t_{\alpha,\,n_1+n_2-2}$$

where $t_{\alpha,\,n_1+n_2-2}$ is a one sided t variate with α level of significance and $n_1 + n_2 - 2$ degrees of freedom and is obtained from "t distribution" tables.

Aspin – Walch test for means of normal distributions when $\sigma_1 \neq \sigma_2$
Aspin – Walch statistic is given by:

$$t = \frac{\overline{x}_1 - \overline{x}_2}{\sqrt{v}}$$

"t" follows a "t distribution" with "f" degrees of freedom

where $\quad v = \dfrac{s_1^2}{n_1} + \dfrac{s_2^2}{n_2} \quad$ and $\quad f = \dfrac{v^2}{\dfrac{(s_1^2/n_1)^2}{n_1 - 1} + \dfrac{(s_2^2/n_2)^2}{n_2 - 1}}$.

Various examples of two sample t test and Aspin – Walch test:

Example: In an experiment, two groups of dogs were weighed following inhalation of a vapour. Data is as follows:

Dog weight with inhalation of vapour (x_1) (kg)	Dog weight without inhalation of vapour i.e. Control group (x_2) (kg)
8.3	8.4
8.8	10.2
9.3	9.6
9.3	9.4

Objective is to test whether the weights of two groups of dogs are same or not.

Solution:

Here the parameter of interest is dog weight, i.e., μ_1= Mean dog weight with inhalation and μ_2= Mean dog weight without inhalation. So null and alternative hypotheses are

$$H_0 : \mu_1 = \mu_2 \quad \text{vs.} \quad H_1 : \mu_1 \neq \mu_2$$

Let mean, sum of squares and variance of dog weight with and without inhalation are $\overline{x}_1$, ss_1 and s_1^2 and $\overline{x}_2$, ss_2 and s_2^2. From the above data

Table: Required calculations for $\overline{x}_1$, ss_1 and s_1^2 and $\overline{x}_2$, ss_2 and s_2^2

Dog weight with inhalation (x_1)	x_1^2	Dog weight without inhalation (Control group) (x_2)	x_2^2
8.3	68.89	8.4	70.56
8.8	77.44	10.2	104.04

	9.3	86.49	9.6	92.16
	9.3	86.49	9.4	88.36
Sum	35.7	319.31	37.6	355.12
Mean	8.92		9.40	

$$\overline{x}_1 = 8.92 \qquad\qquad \overline{x}_2 = 9.40$$

$$n_1 = 4 \qquad\qquad n_2 = 4$$

$$ss_1 = \sum x_i^2 - n_1 * \overline{x}_1^2 = 319.31 - 4*8.92^2 = 1.0444$$

$$ss_2 = \sum x_i^2 - n_2 * \overline{x}_2^2 = 355.12 - 4*9.40^2 = 1.68$$

$$s_1^2 = \frac{ss_1}{(n_1 - 1)} = \frac{1.0444}{3} = 0.3481$$

$$s_2^2 = \frac{ss_2}{(n_2 - 1)} = \frac{1.68}{3} = 0.56$$

As variances are not known but calculated from the samples, first we have to test for equality of variances. As $s_2^2 > s_1^2$, s_2^2 is kept in the numerator.

$$F = \frac{s_2^2}{s_1^2} = \frac{0.56}{0.3481} = 1.6087$$

$$F_{3,3,0.05} = 15.4 \quad \text{(From "F" distribution tables)}$$

As calculated value of F (=1.6087) is less than the tabulated value (=15.4), accept the null hypothesis of equal variances, i.e., the variances of the populations from which the two samples are taken are not different. Hence two sample t test should be applied.

So the pooled variance is

$$s_p^2 = \frac{(n_1 - 1)s_1^2 + (n_2 - 1)s_2^2}{(n_1 - 1) + (n_2 - 1)} = = \frac{1.0444 + 1.68}{(4-1) + (4-1)}$$

$$s_p = 0.6738$$

t statistic is

$$t = \frac{\overline{x}_1 - \overline{x}_2}{s_p \sqrt{\frac{1}{n_1} + \frac{1}{n_2}}} = \frac{8.92 - 9.40}{0.6738 * \sqrt{\frac{1}{4} + \frac{1}{4}}} = \frac{-0.48}{0.6738 * \sqrt{\frac{1}{2}}} = -1.08$$

Degrees of freedom $= n_1 + n_2 - 2 = 4+4-2 = 6$

$t_{6,0.05(2)} = 2.447$ (From "t" distribution tables)

As calculated t (=-1.08) is greater than "-t" tabulated (=-2.447), null hypothesis is accepted, i.e., weight of dogs are same for with and without inhalation groups. This infers that weight of dogs is independent of inhalation.

Example: Two batches of conventional diclofenac sodium tablets (50 mg) were made using (i) starch (ii) cross linked sodium CMC. An experiment was carried out to calculate the amount dissolved in 30 minutes taking a sample of 7 tablets from each batch. Amount dissolved after 30 minutes for each tablet is as follows:

Amount dissolved in 30 min of diclofenac sodium tablets with starch (mg)	Amount dissolved in 30 min of diclofenac sodium tablets with cross linked sodium CMC (mg)
45	38
50	47
47	42
43	39
41	32
49	36
35	41

Test whether the amount dissolved in 30 minutes is same for both the batches.

Solution:
Here the parameter of interest is amount dissolved in 30 minutes, i.e., μ= Mean amount dissolved in 30 minutes. As objective is to test whether the mean amount dissolved in 30 minutes is same for both the batches or not, null and alternative hypotheses are

$$H_0 : \mu_1 = \mu_2 \quad \text{vs.} \quad H_1 : \mu_1 \neq \mu_2$$

where μ_1= Mean amount dissolved in 30 min of diclofenac sodium tablets with starch and μ_2= Mean amount dissolved in 30 min of diclofenac sodium tablets with cross linked sodium CMC.

Let mean & variance of diclofenac sodium tablets with starch and diclofenac sodium tablets with cross linked sodium CMC are $\bar{x}_1$ & s_1^2 and $\bar{x}_2$ & s_2^2. From the above data

$\overline{x}_1 = 44.29$ $\overline{x}_2 = 39.29$

$s_1^2 = 26.9$ $s_2^2 = 22.57$

$n_1 = 7$ $n_2 = 7$

As variances are not known but calculated from the samples, first we have to test for equality of variances.

$$F = \frac{s_1^2}{s_2^2} = \frac{26.9}{22.57} = 1.19$$

$F_{6,6,0.05(2)} = 5.82$ (From "F" distribution tables)

As calculated value of F (=1.19) is less than the tabulated value (=5.82), accept the null hypothesis of equal variances, i.e., the variances of the populations from which the two samples are taken are not different. Hence two sample t test should be applied.

So pooled variance is

$$s_p^2 = \frac{(n_1 - 1)s_1^2 + (n_2 - 1)s_2^2}{(n_1 - 1) + (n_2 - 1)} = \frac{6*26.9 + 6*22.57}{12} = 24.74$$

t statistic is

$$t = \frac{\overline{x}_1 - \overline{x}_2}{s_p \sqrt{\dfrac{1}{n_1} + \dfrac{1}{n_2}}} = \frac{44.29 - 39.29}{\sqrt{24.74 * (\dfrac{1}{7} + \dfrac{1}{7})}} = 1.88$$

Degrees of freedom $= n_1 + n_2 - 2 = 7 + 7 - 2 = 12$

't' at 5% level of significance with 12 df is 2.18 (From "t" distribution tables).

As t calculated (=1.88) is less than t tabulated (=2.18), the amount dissolved in 30 minutes from each batch is not statistically significantly different from each other at 5% level of significance based on samples of 7 tablets from each batch.

Example: Two different immediate release tablet formulations (A and B) of a drug are compared with respect to extent of dissolution. Ten tablets of each formulation were tested, and the amount of drug dissolved at the end of 15 min was calculated. The results are as follows:

Sr.No.	Amount of drug dissolved for A (mg)	Amount of drug dissolved for B (mg)
1	68	74
2	84	71
3	81	79
4	85	63
5	75	80
6	69	61
7	80	69
8	76	72
9	79	80
10	74	65

The objective is to determine if the extent of dissolution differs at 15 min for the two formulations.

Solution:

Here the parameter of interest is amount dissolved in 15 minutes, i.e., μ = Mean amount dissolved in 15 minutes. So null and alternative hypotheses are

$$H_0 : \mu_1 = \mu_2 \quad \text{vs.} \quad H_1 : \mu_1 \neq \mu_2$$

where μ_1 = Mean amount dissolved in 15 min for formulation A and μ_2 = Mean amount dissolved in 15 min for formulation B.

Let mean, variance & sd of formulation A and formulation B are $\bar{x}_1$, s_1^2 & sd_1 and $\bar{x}_2$, s_2^2 & sd_2. From the above data

$$\bar{x}_1 = 77.10 \qquad\qquad \bar{x}_2 = 71.40$$
$$s_1^2 = 33.43 \qquad\qquad s_2^2 = 48.71$$
$$sd_1 = 5.78 \qquad\qquad sd_2 = 6.98$$
$$n_1 = 10 \qquad\qquad n_2 = 10$$

As variances are not known but calculated from the samples, we have to test for equality of variances. As s_2^2 is greater than s_1^2, it is better to take s_2^2 in numerator. So F statistic will be

$$F = \frac{s_2^2}{s_1^2} = \frac{48.71}{33.43} = 1.46$$

$F_{9,9,0.05(2)} = 4.03$ (From "F" distribution tables)

As calculated value of F (=1.46) is less than the tabulated value (=4.03), accept the null hypothesis of equal variances, i.e., the variances of the populations from which the two samples are taken are not different. Hence two sample t test should be applied.

So pooled variance is

$$s_p^2 = \frac{(n_1 - 1)s_1^2 + (n_2 - 1)s_2^2}{(n_1 - 1) + (n_2 - 1)} = \frac{9*33.43 + 9*48.71}{18} = 41.07$$

$$s_p = 6.41$$

Subsequently, t statistic is

$$t = \frac{\bar{x}_1 - \bar{x}_2}{s_p \sqrt{\dfrac{1}{n_1} + \dfrac{1}{n_2}}} = \frac{77.10 - 71.40}{6.41 * \sqrt{\dfrac{1}{10} + \dfrac{1}{10}}} = 1.99$$

Degrees of freedom = $n_1 + n_2 - 2 = 10 + 10 - 2 = 18$

't' value at the 5% level of significant with 18 df is 2.101 (From "t" distribution tables).

As t calculated (=1.99) is less than t tabulated (=2.101), the difference in the mean amount dissolved in 15 minutes taking a sample of 10 tablets from each formulation is not statistically significant at 5% level of significance. Therefore, the two formulations have the same extent of dissolution at 15 minute.

Example: One group of wistar rats was fed with water ad libitum and was mated with their corresponding male to obtain F1 generation and breeding continued to obtain the F5 generation. This was treated as the control group. Another group of animals were fed on 6% alcohol ad libitum instead of water. Breeding was done in the same pattern to obtain the F5 generation. Pups' were weighed and results are given in the following table. The aim was to determine whether the weights of pups of alcohol fed mothers was significantly different from that of water fed mothers.

Pup's weight of alcohol fed mothers (gm)	Pup's weight of water fed mother (gm)
7	10
6.5	12
6.0	15

8.0	11
7.5	12.5
5.0	14
8.0	10.5
6.5	13
5.5	10.5
5.0	12
6.0	13.5
8.0	10.0
7.0	11.0
8.0	12.0
9.0	11.0

Solution:

Here the parameter of interest is pup's weight, i.e., μ= Mean pup's weight. Also our interest is in testing equality versus not equality, so null and alternative hypotheses are

$$H_0 : \mu_1 = \mu_2 \quad \text{vs.} \quad H_1 : \mu_1 \neq \mu_2$$

where μ_1= Mean pup's weight of alcohol fed mothers and μ_2= Mean pup's weight of water fed mothers.

Let mean & variance of pup's weight of alcohol fed mothers and of water fed mothers are $\bar{x}_1$ & s_1^2 and $\bar{x}_2$ & s_2^2. From the above data

$$\bar{x}_1 = 6.87 \qquad \bar{x}_2 = 11.87$$
$$s_1^2 = 1.481 \qquad s_2^2 = 2.267$$
$$n_1 = 15 \qquad n_2 = 15$$

As variances are not known but calculated from the samples, first we have to test for equality of variances. As $s_2^2 > s_1^2$,

$$F = \frac{s_2^2}{s_1^2} = \frac{2.267}{1.481} = 1.531$$

$$F_{14,14,0.05} = 2.98 \qquad \text{(From "F" distribution tables)}$$

As calculated value of F (=1.531) is less than the tabulated value (=2.98), accept the null hypothesis of equal variances, i.e., the variances of the populations from which the two samples are taken are not different. Hence two sample t test should be applied.

So pooled standard deviation is

$$s_p = \sqrt{\frac{(n_1-1)s_1^2 + (n_2-1)s_2^2}{n_1 + n_2 - 2}}$$

$$= \sqrt{\frac{(15-1)(1.481)+(15-1)(2.267)}{15+15-2}} = \sqrt{\frac{52.467}{28}} = \sqrt{1.874} = 1.369$$

$$t = \frac{\overline{x_1} - \overline{x_2}}{s_p\sqrt{\frac{1}{n_1}+\frac{1}{n_2}}} = \frac{6.87-11.87}{1.369*\sqrt{\frac{1}{15}+\frac{1}{15}}} = \frac{-5.0}{1.369*\sqrt{\frac{2}{15}}} = -10.06$$

At α = 0.05 and degree of freedom = $n_1 + n_2 - 2$ = 28, t = 2.048 from "t" distribution table.
As t(calculated) < -t(tabulated), we reject the null hypothesis of equality.
Hence it can be concluded that weight of pups of water fed mothers is statistically significantly greater than that of alcohol fed mothers.

Example: An experiment was carried out to test the effect of gelling agents gelrite and agar on the fresh and dry weight of D.innoxia callus after 4 weeks of culture. Fresh and dry weight of D.innoxia callus (n=5) after getting two agents are given below:

Fresh wt		Dry wt	
Gelrite	**Agar**	**Gelrite**	**Agar**
26.362	21.136	0.7562	0.6600
25.646	18.884	0.7357	0.5635
25.567	22.440	0.6821	0.6414
24.024	20.949	0.7058	0.6254
23.529	24.274	0.7347	0.6409

Solution:
Here the parameters of interest are fresh and dry weight of D.innoxia callus after 4 weeks of culture, i.e., μ= Mean fresh weight of D.innoxia callus after 4 weeks

of culture or mean dry weight of D.innoxia callus after 4 weeks of culture. We will consider fresh and dry weight cases separately.

Fresh weight:

Null and alternative hypotheses are

$$H_0 : \mu_1 = \mu_2 \quad \text{vs.} \quad H_1 : \mu_1 \neq \mu_2$$

where μ_1= Mean fresh weight of D.innoxia callus after 4 weeks of culture with gelling agents gelrite and μ_2= Mean fresh weight of D.innoxia callus after 4 weeks of culture with gelling agents agar.

Let mean and variance of fresh weight of D.innoxia callus after 4 weeks of culture getting gelling agents gelrite and agar are $\bar{x}_1$ and s_1^2 and $\bar{x}_2$ and s_2^2. From the above data

$$\bar{x}_1 = 25.025 \qquad\qquad \bar{x}_2 = 21.537$$
$$s_1^2 = 1.427 \qquad\qquad s_2^2 = 3.963$$
$$n_1 = 5 \qquad\qquad n_2 = 5$$

As variances are not known but calculated from the samples, first we have to test for equality of variances. As $s_2^2 > s_1^2$,

$$F = \frac{s_2^2}{s_1^2} = \frac{3.963}{1.427} = 2.78$$

At 5% level of significance, $F_{4,4,0.05} = 9.60$ (From "F" distribution tables)

As calculated value of F (=2.78) is less than the tabulated value (=9.60), accept the null hypothesis of equal variances, i.e., the variances of the populations from which the two samples are taken are not different. Therefore, two sample t test should be applied.

So pooled standard deviation is

$$s_p = \sqrt{\frac{(n_1 - 1)s_1^2 + (n_2 - 1)s_2^2}{n_1 + n_2 - 2}} = \sqrt{\frac{(5-1)*1.427 + (5-1)*3.963}{5+5-2}} = \sqrt{2.6948}$$

$$= 1.6416$$

Subsequently, t statistic is

$$t_{cal} = \frac{\overline{x}_1 - \overline{x}_2}{s_p\sqrt{\dfrac{1}{n_1} + \dfrac{1}{n_2}}} = \frac{25.025 - 21.537}{1.6416\sqrt{\dfrac{1}{5} + \dfrac{1}{5}}} = \frac{3.489}{1.038} = 3.361$$

At 5% level of significance and 8 (=5+5-2) degrees of freedom, t is 2.306 from "t" distribution tables.

Since t calculated (=3.361) is greater than t tabulated (2.306), null hypothesis is rejected i.e., there is an effect of agents gelrite and agar on the fresh weight of D.innoxia callus after 4 weeks of culture.

<u>Dry weight</u>

Null and alternative hypotheses are

$$H_0 : \mu_1 = \mu_2 \quad \text{vs.} \quad H_1 : \mu_1 \neq \mu_2$$

where μ_1= Mean dry weight of D.innoxia callus after 4 weeks of culture with gelling agents gelrite and μ_2= Mean dry weight of D.innoxia callus after 4 weeks of culture with gelling agents agar.

Let mean & variance of dry weight of D.innoxia callus after 4 weeks of culture getting gelling agents gelrite and agar are $\overline{x}_1$ & s_1^2 and $\overline{x}_2$ & s_2^2. From the above data

$$\overline{x}_1 = 0.723 \qquad\qquad \overline{x}_2 = 0.626$$
$$s_1^2 = 0.0008 \qquad\qquad s_2^2 = 0.0014$$
$$n_1 = 5 \qquad\qquad n_2 = 5$$

As variances are not known but calculated from the samples, first we have to test for equality of variances. As $s_2^2 > s_1^2$,

$$F = \frac{s_2^2}{s_1^2} = \frac{0.0014}{0.0008} = 1.639$$

At 5% level of significance, $F_{4,4,0.05} = 9.60$ (From "F" distribution tables)

As calculated value of F (1.639) is less than the tabulated value (9.60) , accept the null hypothesis of equal variances, i.e., the variances of the population from which the two samples are taken are not different. Therefore, two sample t test should be applied.

So pooled standard deviation is

$$s_p = \sqrt{\frac{(n_1-1)s_1^2 + (n_2-1)s_2^2}{n_1+n_2-2}} = \sqrt{\frac{(5-1)*.0008 + (5-1)*0.0014}{5+5-2}} = \sqrt{0.0011}$$

$$= 0.03312$$

Subsequently, t statistic is

$$t_{cal} = \frac{\overline{x}_1 - \overline{x}_2}{s_p\sqrt{\dfrac{1}{n_1}+\dfrac{1}{n_2}}} = \frac{0.723 - 0.626}{0.03312\sqrt{\dfrac{1}{5}+\dfrac{1}{5}}} = \frac{0.097}{0.021} = 4.62$$

At 5% level of significance and 8 (=5+5-2) degrees of freedom, t is 2.306 from "t" distribution tables.

Since t calculated (=4.62) is greater than t tabulated value (2.306), null hypothesis is rejected i.e., there is an effect of agents gelrite and agar on the dry weight of D.innoxia callus after 4 weeks of culture.

Hence it can be concluded that fresh as well as dry weight of D.innoxia callus was more when gelling agent was gelrite.

Example: Na-I131 was mixed in polymer solution and coated on nonpareil seeds. Objective was to assess the stability of Na-I131 in pellet dosage form on exposure to high humidity. Hence the coated pellets (n=5) were kept at 250°C and 250°C + 75% RH, for a period of 6 months. Then iodide iron was analyzed using an I-selective electrode. Iodide iron content (mg/10 gms) of coated pellets are given as mean (± sd).

Iodide iron content (mg/10 gms) found after 6 months storage at two different conditions	
Condition	
250°C	**250°C + 75%RH**
98.0 ((± 2.9)	95.7 ((± 2.7)

Test whether Na-I131 in pellet dosage is stable or not on exposure to high humidity.

Solution:
As iodide iron content was the parameter to see the effect of humidity, the parameters of interest is iodide iron content, i.e., μ = Mean iodide iron content.

μ_1 = Mean iodide iron content at 250°C, μ_2 = Mean iodide iron content at 250°C + 75% RH. We have to test whether Na-I131 in pellet dosage is stable or not on exposure to high humidity, i.e., iodide iron content with 75% RH humid condition is less than or more than the iodide iron content without this condition. Hence null and alternative hypotheses are

$$H_0 : \mu_1 = \mu_2 \quad \text{vs.} \quad H_1 : \mu_1 \neq \mu_2$$

From the given data

$$\overline{x}_1 = 98.0 \qquad\qquad \overline{x}_2 = 95.7$$
$$s_1 = 2.9 \qquad\qquad s_1 = 2.7$$
$$s_1^2 = 8.41 \qquad\qquad s_2^2 = 7.29$$
$$n_1 = 5 \qquad\qquad n_2 = 5$$

As variances are not known but calculated from the samples, first we have to test for equality of variances.

$$F = \frac{s_1^2}{s_2^2} = \frac{8.41}{7.29} = 1.15$$

At 5% level of significance, $F_{4,4,0.05} = 9.60$ (From "F" distribution tables)

As calculated value of F (=1.15) is less than the tabulated value (9.60), accept the null hypothesis of equal variances, i.e., the variances of the populations from which the two samples are taken are not different. Therefore, two sample t test should be applied

So pooled standard deviation is

$$s_p = \sqrt{\frac{(n_1 - 1)s_1^2 + (n_2 - 1)s_2^2}{n_1 + n_2 - 2}} = \sqrt{\frac{(5-1)*8.41 + (5-1)*7.29}{5 + 5 - 2}} = \sqrt{7.85}$$

$$= 2.8018$$

Subsequently, t statistic is

$$t_{cal} = \frac{\overline{x}_1 - \overline{x}_2}{s_p\sqrt{\dfrac{1}{n_1} + \dfrac{1}{n_2}}} = \frac{98.0 - 95.7}{2.8\sqrt{\dfrac{1}{5} + \dfrac{1}{5}}} = \frac{2.3}{1.772} = 1.298$$

At 5% level of significance and 8 (=5+5-2) degrees of freedom, t is 2.306 from "t" distribution tables.

Since t calculated (=1.298) is less than t tabulated (2.306), null hypothesis is accepted. It seems that storage at high humidity does not affect iodide iron content of coated pellets and hence Na-I131 in pellet dosage form is stable on exposure to high humidity.

Example: Ciprofloxacin raw material contains a degradation product, analog A. Solution dosage form of ciprofloxacin (eye drops) was prepared using this raw material. The amount of analog A was estimated in the raw material and from eye drops, stored at accelerated condition of temperature (450C, 3months). The data obtained is as follow:

Amount of analog A in raw material: 0.0388, 0.0343, 0.0398, 0.0388, 0.0371, 0.0355 and 0.0368 mg
Amount of analog A in Eye drops: 0.0409, 0.0409, 0.0424, and 0.0434 mg

The objective of the experiment was to check whether amount of analog A was same in raw material and eye drops.

Solution:
Here the parameters of interest is amount of analog A, i.e., μ= Mean amount of analog A.
μ_1= Mean amount of analog A in raw material, μ_2 = Mean amount of analog A in eye drops. We have to test whether mean amount of analog A is same in raw material and eye drops. Hence null and alternative hypotheses are

$$H_0 : \mu_1 = \mu_2 \quad \text{vs.} \quad H_1 : \mu_1 \neq \mu_2$$

From the given data

$$\bar{x}_1 = 0.0373 \qquad\qquad \bar{x}_2 = 0.0419$$
$$s_1 = 0.00197 \qquad\qquad s_2 = 0.00123$$
$$n_1 = 7 \qquad\qquad n_2 = 4$$

As variances are not known but calculated from the samples, first we have to test for equality of variances. As $s_2^2 < s_1^2$,

$$F = \frac{s_1^2}{s_2^2} = \frac{0.00197^2}{0.00123^2} = 2.57$$

At 5% level of significance, $F_{6,3,0.05} = 14.7$ \qquad\qquad (From "F" distribution tables)

As calculated value of F (=2.57) is less than the tabulated value (= 14.7), accept the null hypothesis of equal variances, i.e., the variances of the populations from which the two samples are taken are not different. Therefore, two sample t test should be applied

So pooled standard deviation is

$$s_p = \sqrt{\frac{(n_1-1)s_1^2 + (n_2-1)s_2^2}{n_1+n_2-2}} = \sqrt{\frac{(7-1)*0.00197^2 + (4-1)*0.00123^2}{7+4-2}}$$

$$= 0.001757$$

Subsequently, t statistic is

$$t_{cal} = \frac{\bar{x}_1 - \bar{x}_2}{s_p\sqrt{\frac{1}{n_1}+\frac{1}{n_2}}} = \frac{0.0373-0.0419}{0.001757\sqrt{\frac{1}{7}+\frac{1}{4}}} = \frac{-0.0046}{0.00110} = -4.1773$$

At 5% level of significance and 9 (=7+4-2) degrees of freedom, t is 2.262 from "t" distribution tables.

Since t calculated (-4.1773) is less than "-t" tabulated (-2.262), null hypothesis is rejected. It implies that mean amount of analog A is not same in raw material and eye drops.

Example: An experiment was conducted to see the effect of certain treatment on 10 morphological parameters of leaf. Of which one was leaf length. There was one control group and one treatment group consisting of 25 units. Leaf length (in cm) was measured in control as well as treated group. Results are given in the following table.

Leaf length of Control group (cm)	10.7	10.5	13.2	5	11.6	9	12	11.5	7.6	6.5
	8	10	8.3	6.5	8.5	12.2	6	15	3	7.8
	7	6.5	6.8	7.5	8.6					
Leaf length of Treated group (cm)	10.5	12.8	8.5	9.2	7.5	9.5	6.6	14	7.6	9.2
	9	11.5	9	13.2	11.7	11	10.6	10.2	12	12.5
	14.2	9	8	10.6	11.5					

Test whether the treatment increases the leaf length.

Solution: Here the parameter of interest is leaf length, i.e., μ = Average leaf length. μ_1 = Average leaf length for control group, μ_2 = Average leaf length for treatment group

As the interest is to test whether the treatment increases the leaf length, null and alternative hypotheses are

$$H_0 : \mu_1 = \mu_2 \quad \text{vs.} \quad H_1 : \mu_1 < \mu_2$$

From the given data

$\bar{x}_1 = 8.772$	$\bar{x}_2 = 10.376$
$s_1^2 = 7.786$	$s_2^2 = 4.258$
$n_1 = 25$	$n_2 = 25$

As variances are not known but calculated from the samples, first we have to test for equality of variances.

$$F = \frac{s_1^2}{s_2^2} = \frac{7.786}{4.258} = 1.83$$

At 5% level of significance, $F_{24,24,0.05} = 2.27$ (From "F" distribution tables)

As calculated value of F (=1.83) is less than the tabulated value (=2.27), accept the null hypothesis of equal variances, i.e., the variances of the populations from which the two samples are taken are not different. Therefore, two sample t test should be applied.

So pooled standard deviation is

$$s_p = \sqrt{\frac{(n_1-1)s_1^2 + (n_2-1)s_2^2}{n_1 + n_2 - 2}} = \sqrt{\frac{(25-1)*7.786 + (25-1)*4.258}{25+25-2}} = \sqrt{6.022}$$

$$= 2.45$$

Subsequently, t statistic is

$$t_{cal} = \frac{\bar{x}_1 - \bar{x}_2}{s_p\sqrt{\frac{1}{n_1} + \frac{1}{n_2}}} = \frac{8.772 - 10.376}{2.45\sqrt{\frac{1}{25} + \frac{1}{25}}} = \frac{-1.604}{0.694} = -2.311$$

At 5% level of significance and 48 (=25+25-2) degrees of freedom, t is 2.011 from "t" distribution tables. Since t calculated is less than "-t(tabulated)", null hypothesis is rejected. It is concluded that leaf length of treated group is more than the control group and hence treatment increases the leaf length.

Example: Following data gives the amount of a compound in market and tissue culture samples:

Compound in market sample: 0.073, 0,075, 0.077, 0.073, 0.074, 0.072

Compound in tissue culture sample : 0.032, 0.035, 0.037, 0.035, 0.038, 0.031

Test whether amount of compound is same in market and tissue culture samples.

Solution:

Here the parameters of interest is amount of compound, i.e., μ= Mean amount of compound.

μ_1= Mean amount of compound in market samples, μ_2 = Mean amount of compound in tissue culture samples. We have to test whether amount of compound is same in market samples and tissue culture samples. Hence null and alternative hypotheses are

$$H_0 : \mu_1 = \mu_2 \quad \text{vs.} \quad H_1 : \mu_1 \neq \mu_2$$

From the given data

$$\overline{x}_1 = 0.075 \qquad\qquad \overline{x}_2 = 0.035$$
$$s_1 = 0.0018 \qquad\qquad s_2 = 0.0027$$
$$n_1 = 6 \qquad\qquad n_2 = 6$$

As variances are not known but calculated from the samples, first we have to test for equality of variances. As $s_2^2 > s_1^2$,

$$F = \frac{s_2^2}{s_1^2} = \frac{0.0027^2}{0.0018^2} = 2.33$$

At 5% level of significance, $F_{5,5,0.05} = 7.15$ \qquad (From "F" distribution tables)

As calculated value of F (=2.33) is less than the tabulated value (= 7.15), accept the null hypothesis of equal variances, i.e., the variances of the populations from which the two samples are taken are not different. Therefore, two sample t test should be applied.

So pooled standard deviation is

$$s_p = \sqrt{\frac{(n_1-1)s_1^2 + (n_2-1)s_2^2}{n_1 + n_2 - 2}} = \sqrt{\frac{(6-1)*0.0018^2 + (6-1)*0.0027^2}{6+6-2}}$$

$$= 0.0023$$

Subsequently t statistic is

$$t_{cal} = \frac{\overline{x}_1 - \overline{x}_2}{s_p \sqrt{\dfrac{1}{n_1} + \dfrac{1}{n_2}}} = \frac{0.075 - 0.035}{0.0023 \sqrt{\dfrac{1}{6} + \dfrac{1}{6}}} = \frac{-0.04}{0.0013} = -30$$

At 5% level of significance and 10 (=6+6-2) degrees of freedom, t is 2.228 from "t" distribution tables and hence –t(tabulated) is -2.228.

Since t calculated (-30) is less than -2.228, null hypothesis is rejected. It implies that amount of compound is not same in market samples and tissue culture samples.

Example: An experiment was conducted to measure the total phenolics in shoots grown in vitro and in vivo. Following table gives the amount of phenolics in shoots grown *in vitro* and *in vivo*:

in vitro	*in vivo*
0.318	0.628
0.291	0.640
0.331	0.601

Test whether amount of phenolics in shoots grown in vitro and in vivo are equal.

Solution:

Here parameter of interest is amount of phenolics, i.e., μ= Mean amount of phenolics.

μ_1= Mean amount of phenolics in shoots grown *in vitro*, μ_2 = Mean amount of phenolics in shoots grown *in vivo*. We have to test whether amount of phenolics is same in shoots grown *in vitro* and *in vivo*. Hence null and alternative hypotheses are

$$H_0 : \mu_1 = \mu_2 \quad \text{vs.} \quad H_1 : \mu_1 \neq \mu_2$$

From the given data

$\overline{x}_1 = 0.313$ $\qquad\qquad$ $\overline{x}_2 = 0.623$

$s_1 = 0.0204$ $\qquad\qquad$ $s_2 = 0.02$

$n_1 = 3$ $\qquad\qquad$ $n_2 = 3$

As variances are not known but calculated from the samples, first we have to test for equality of variances.

$$F = \frac{s_1^2}{s_2^2} = \frac{0.0204^2}{0.02^2} = 1.04$$

At 5% level of significance, $F_{2,2,0.05} = 39.0$ (From "F" distribution tables)

As calculated value of F (=1.04) is less than the tabulated value (= 39.0), accept the null hypothesis of equal variances, i.e., the variances of the populations from which the two samples are taken are not different. Therefore, two sample t test should be applied.

So pooled standard deviation is

$$s_p = \sqrt{\frac{(n_1-1)s_1^2 + (n_2-1)s_2^2}{n_1+n_2-2}} = \sqrt{\frac{(3-1)*0.0204^2 + (3-1)*0.02^2}{3+3-2}}$$

$= 0.0202$

Subsequently, t statistic is

$$t_{cal} = \frac{\overline{x}_1 - \overline{x}_2}{s_p\sqrt{\frac{1}{n_1}+\frac{1}{n_2}}} = \frac{0.313-0.623}{0.0202\sqrt{\frac{1}{3}+\frac{1}{3}}} = \frac{-0.31}{0.0165} = -18.8$$

At 5% level of significance and 4 (=3+3-2) degrees of freedom, t is 2.776 from "t" distribution tablesand, hence –t(tabulated) will be -2.776.

Since t(calculated) (=-18.8) is less than -2.776, null hypothesis is rejected. It implies that mean amount of phenolics in shoots grown *in vivo* is not same as mean amount of phenolics in shoots grown *in vitro*.

Example: The below table shows lipid peroxidation (LPO; nmol/g fr. wt./min) in the uterus of treated Swiss albino mice on day 1 and day 3 (4:40 am) of pregnancy. Test whether there is any change in LPO levels in day 1 and day 3 of pregnancy of rats.

Rabbit No.	LPO level in the uterus of Swiss albino mice on Day 1 of one group (x_{1i})	LPO level in the uterus of Swiss albino mice on Day 3 of other group (x_{2i})
1	3.05	1.79
2	2.91	1.36
3	4.07	1.5
4	4.53	1.71

Solution: Here the parameters of interest is LPO levels in the uterus of treated Swiss albino mice, i.e., μ= Mean LPO levels in the uterus of treated Swiss albino mice.

μ_1= Mean LPO levels in the uterus of treated Swiss albino mice on day 1, μ_2 = Mean LPO levels in the uterus of treated Swiss albino mice on day 3. We have to test whether LPO levels in the uterus of treated Swiss albino mice on day 1 is same as LPO levels in the uterus of treated Swiss albino mice on day 3. Hence null and alternative hypotheses are

$$H_0 : \mu_1 = \mu_2 \quad \text{vs.} \quad H_1 : \mu_1 \neq \mu_2$$

From the given data

$$\bar{x}_1 = 3.64 \qquad\qquad \bar{x}_2 = 1.59$$

$$s_1 = 0.7870 \qquad\qquad s_2 = 0.1961$$

$$s_1^2 = 0.6193 \qquad\qquad s_2^2 = 0.0385$$

$$n_1 = 4 \qquad\qquad n_2 = 4$$

As variances are not known but calculated from the samples, first we have to test for equality of variances.

$$F = \frac{s_1^2}{s_2^2} = \frac{0.6193}{0.0385} = 16.10$$

At 5% level of significance, $F_{3,3,0.05(2)} = 15.4$ \qquad (From "F" distribution tables)

As calculated value of F (=16.10) is more than the tabulated value (15.4), reject the null hypothesis of equal variances, i.e., the variances of the populations from which the two samples are taken are different. So instead of two sample t test, Aspin -Walch test should be applied.

$$v = \frac{s_1^2}{n_1} + \frac{s_2^2}{n_2} = \frac{0.6193}{4} + \frac{0.0385}{4} = 0.1645$$

Aspin – Walch statistic is given by:

$$t = \frac{\bar{x}_1 - \bar{x}_2}{\sqrt{v}} = \frac{3.64 - 1.59}{\sqrt{0.1645}} = \frac{2.05}{0.4055} = 5.06$$

degrees of freedom

$$f = \frac{v^2}{\dfrac{(s_1^2/n_1)^2}{n_1 - 1} + \dfrac{(s_2^2/n_2)^2}{n_2 - 1}} = \frac{0.1645^2}{\dfrac{(0.6193/4)^2}{4-1} + \dfrac{(0.0385/4)^2}{4-1}} = 3.37$$

As t follows a "t distribution" with f (=3) degrees of freedom, $t_{3,0.05(2)}$ is 3.182 from "t" distribution tables.

Since t calculated (5.06) is more than t tabulated (3.182), null hypothesis is rejected. It implies that mean LPO levels in the uterus of treated Swiss albino mice on day 1 is different than mean LPO levels in the uterus of treated Swiss albino mice on day 3.

PAIRED SAMPLE CASE: PAIRED-T TEST

- Sometimes there are situations when observations are taken on two occasions or from two sites of the same subject / unit. For example, study of a new diet program, the weights before and after the program yields paired observations.
- Basic purpose of pairing is to remove the possible extraneous factors that may cause a difference in outcome.
- Sometimes pairing is created by the investigator who matches subjects or items on important characteristics so that members of a pair are as alike as possible with regard to those characteristic.
- Consequently, observations are not independent, hence two sample t test can not be applied as there is one assumption of observations being independent for applying two sample t test. For these cases, paired t test is applied.

Suppose we have two sets of observations $(x_{11}, x_{12}, ..., x_{1n})$ and $(x_{21}, x_{22}, ..., x_{2n})$ taken on n individuals / experimental units. These two sets of observations can be observations on two occasions i.e. pre and post treatment or observations after applying two treatment on left and right sites of an individual / experimental unit. On this data, paired t test is applied as follows:

1. Take the deviation of observations for each pair of observations as follows:

Subject / unit No. (i)	1st observation (x_{1i})	2nd observation (x_{2i})	$d_i = x_{1i} - x_{2i}$
1	x_{11}	x_{21}	$x_{11} - x_{21}$
2	x_{12}		$x_{12} - x_{22}$
.	.	.	.
.	.	.	.
n-1	$x_{1(n-1)}$	$x_{2(n-1)}$	$x_{1(n-1)} - x_{2(n-1)}$
n	x_{1n}	x_{2n}	$x_{1n} - x_{2n}$
		Sum	$\sum_{i=1}^{n} d_i$
		Average	$\bar{d}$

2. Once deviations are calculated, the two sample problem reduces to one sample problem.

3. Now the null hypothesis to be tested in this case may be one of the following:

$H_0: \mu = \Delta$ vs. $H_1: \mu \neq \Delta$,

$H_0: \mu \geq \Delta$ vs. $H_1: \mu < \Delta$,

$H_0: \mu \leq \Delta$ vs. $H_1: \mu > \Delta$,

where μ is the average deviations of two observations taken on an individual / experimental unit of population and Δ can assume any value based on the acceptable difference between the two sets of observations.

4. If data follows a normal distribution, apply one sample t test as follows:

$$t = \frac{\bar{d} - \Delta}{s/\sqrt{n}}$$

where $\bar{d}$ is the mean of d_i and s is the standard deviation of d_i.

5. Depending on the hypotheses, null hypothesis will be rejected if

$$t = \frac{\bar{d} - \Delta}{s/\sqrt{n}} > t_{\alpha/2,n-1} \text{ or } t = \frac{\bar{d} - \Delta}{s/\sqrt{n}} < -t_{\alpha/2,n-1} \text{ for } H_0: \mu = \Delta \text{ vs } H_1: \mu \neq \Delta,$$

$$t = \frac{\bar{d} - \Delta}{s/\sqrt{n}} < -t_{\alpha,n-1} \qquad\qquad \text{for} \qquad H_0: \mu \geq \Delta \text{ vs. } H_1: \mu < \Delta,$$

$$t = \frac{\bar{d} - \Delta}{s/\sqrt{n}} > t_{\alpha,n-1} \qquad\qquad \text{for} \qquad H_0: \mu \leq \Delta \text{ vs } H_1: \mu > \Delta,$$

Example: In an pharmacokinetic experiment, formulation A and formulation B were given on two occasions to 6 animals and blood samples were taken on both the occasions. Pharmacokinetic paramenter AUCs for both the formulations are as follows:

Animal No.	AUC for Formulation (ng.hr/ml) (A)	AUC for Formulation (ng.hr/ml) (B)
1	136	166
2	168	184
3	160	193
4	94	105
5	200	198
6	174	197

Test whether the pharmacokinetic parameter AUC is same for formulations A and B.

Solution:

Here our interest is in area under the curve (AUC) and we want to test whether the AUCs are same for formulations A and B i.e. we want to test if difference between two AUCs is 0 or not. As AUCs were obtained on the same animal after giving treatment A and B on two occasions, we will apply paired t test instead of 2 sample t test. Hence the null hypothesis to be tested in this case will be:

$$H_0: \mu = \Delta \text{ vs. } H_1: \mu \neq \Delta$$

where μ is mean difference in AUC and $\Delta = 0$

First calculate the difference in AUCs for each animal after treatment A and B as follows:

Table: Required calculations for paired t test

Animal	AUC for Formulation (A)	AUC for Formulation (B)	$d_i = $ B-A
1	136	166	30
2	168	184	16
3	160	193	33
4	94	105	11
5	200	198	-2
6	174	197	23
		Mean ($\bar{d}$)	18.5
		sd (s)	13.0

The test statistic will be

$$t = \frac{\bar{d} - \Delta}{s/\sqrt{n}} = \frac{18.5 - 0}{13.0/\sqrt{6}} = \frac{18.5}{5.307} = 3.486$$

$$t_{0.05(2),5} = 2.57 \qquad \text{(From "t" distribution tables)}$$

As t calculated (=3.486) is greater than t tabulated (=2.57), reject the null hypothesis of no difference between AUCs of two formulations. This infers that AUCs for the two formulations are different.

Example: For testing the efficacy of a unani formulation against insomnia, a double blind study in 10 volunteers was carried out. Volunteers were given a high dose of coffee in one phase and the same dose of coffee and drug in the second phase. The latency of sleep was recorded in minutes. Our aim was to determine whether the latency of sleep was significantly reduced after drug.

Solution:

Here the parameter of interest is latency of sleep. Hence μ is the average difference in latency of sleep after high dose of coffee in one phase and the same dose of coffee and drug in the second phase and $\Delta=0$ because our interest is if there is any difference in latency of sleep. As our aim is to determine whether the latency of sleep is significantly more when only high dose of coffee is given, the null and alternative hypotheses to be tested in this case will be:

$$H_0: \mu \leq 0 \qquad \text{vs.} \qquad H_1: \mu > 0$$

So we calculate the deviations in latency of sleep after high dose of coffee in one phase and the same dose of coffee and drug in the second phase as given below:

Table: Required calculations for paired t test

Volunteer No.	Sleep latency after high dose of coffee (min) (x_{1i})	Sleep latency after the same dose of coffee and drug (min) (x_{2i})	$d_i = x_{1i} - x_{2i}$
1	60	22	38
2	56	20	36
3	58	25	33
4	60	15	45
5	55	20	35
6	60	20	40
7	58	18	40
8	58	18	40
9	60	20	40
10	55	20	35
		Mean ($\bar{d}$)	38.20
		sd (s)	3.52

$$t = \frac{\bar{d} - \Delta}{s/\sqrt{n}} = \frac{38.2 - 0}{3.52/\sqrt{10}} = \frac{38.2}{3.52/3.16} = 34.305$$

At 5% level of significance i.e., $\alpha = 0.05(1)$, degrees of freedom $= n-1 = 9$, $t = 1.833$ (From "t" distribution tables).

As t(calculated) (=34.305) > t(tabulated) =(1.833), we reject the null hypothesis that sleep latency after high dose of coffee is same as the sleep latency after high dose of coffee and drug. This infers that sleep latency is more after high dose of coffee only and hence drug helps against insomnia.

Example: Viscosity measurements were done at the start of the experiment and after one month storage at 370°C during stability studies of a liquid paraffin O/W emulsion and are given below.

Sample	Initial viscosity	Viscosity after one month storage at 370°C
1	100.8	100.01
2	101.9	99.3
3	98.7	100.0
4	101.3	101.4
5	102.5	101.9

Objective of the experiment was to check the stability of this liquid paraffin O/W emulsion after one month storage at 370°C.

Solution:
Here parameter of interest is change in viscosity i.e. μ is the average difference in viscosity at two time points. We want to test whether viscosity of liquid paraffin O/W emulsion changes after one month storage at 370°C after, so the null hypothesis will be

$H_0: \mu = 0$ vs. $H_1: \mu \neq 0$

Table: Required calculations for paired t test

Sample	Initial viscosity (x_{1i})	Viscosity after one month storage at 370°C (x_{2i})	$d_i = x_{1i} - x_{2i}$
1	100.8	100.01	0.7
2	101.9	99.3	2.6
3	98.7	100.0	-1.3
4	101.3	101.4	-0.1
5	102.5	101.9	0.6
		Mean ($\bar{d}$)	0.50
		sd (s)	1.42

$$t = \frac{\overline{d} - \Delta}{s/\sqrt{n}} = \frac{\overline{d} - 0}{s}\sqrt{n}$$

$$= \frac{0.5 - 0}{1.42}\sqrt{5}$$

$$= 0.788$$

t = 2.776 at α=0.05(2) and 4 df from "t" distribution tables.

As t(calculated) (0.788) < t(tabulated) (=2.776), null hypothesis is accepted i.e. there is no statistically significant difference in viscosity at two occasions at 5% level of significance. Hence solution is stable up to one month at 370°C.

Example: Fifty (50) μl of phenylephrine HCl and phenylephrine oxazolidinedione (a produrg) solution were instilled in to the left and right eye of 12 New Zealand albino rabbits, respectively. Pupil diameters were measured at 1h after dosing. Increase in pupil diameter after instilling phenylephrine oxazolidinedione and phenylephrine HCl are as follows:

Rabbit No.	1	2	3	4	5	6	7	8	9	10	11	12
Increase in pupil diameter after instilling Phenylephrine oxazolidinedione	53	38	69	57	46	39	73	48	73	74	60	78
Increase in pupil diameter after instilling Phenylephrine HCl	44	40	61	52	32	44	70	41	67	72	53	72

Objective was to assess whether phenylephrine oxazolidinedione possesses better mydriatic activity than phenylephrine HCl.

Solution:
Here our interest is in mydriatic activity which is measured by Pupil diameter i.e. μ is the average difference in pupil diameters after instilling prodrug solution and phenylephrine HCl in right and left eye of rabbits. As we want to test whether phenylephrine oxazolidinedione possesses better mydriatic activity than phenylephrine HCl, our hypotheses will be:

$$H_0: \mu \leq 0 \qquad \text{vs.} \qquad H_1: \mu > 0$$

Increase in pupil diameter after instilling prodrug solution and phenylephrine HCl in right and left eye of rabbits and their deviations are as follows:

Table: Required calculations for paired t test

Rabbit No.	Increase in pupil diameter after instilling Phenylephrine oxazolidinedione (x_{1i})	Increase in pupil diameter after instilling Phenylephrine HCl (x_{2i})	d_i ($x_{1i} - x_{2i}$)
1	53	44	+9
2	38	40	-2
3	69	61	+8
4	57	52	+5
5	46	32	+14
6	39	44	-5
7	73	70	+3
8	48	41	+7
9	73	67	+6
10	74	72	+2
11	60	53	+7
12	78	72	+6
		Mean ($\bar{d}$)	5
		sd (s)	5.027

$$t = \frac{(\bar{d} - 0)\sqrt{n}}{s} = \frac{5\sqrt{12}}{5.03} = 3.445$$

df = n-1 = 12-1 = 11

$t_{11, 0.05(1)} = 1.796$ from "t" distribution tables.

As t_{cal} (=3.445) > t_{tab} (=1.796), H_0 is rejected. This infers that phenylephrine oxazolidinedione (a prodrug) has better mydriatic activity than phenylephrine HCl.

Example: An experiment was conducted to study the effect of 1 month storage at 45°C and 75% RH on weight of rifampicin tablets. Sample for this experiment consists of 11 tablets. Weights (in mg) of tablets were taken at the start of the experiment and after one month storage at 45°C and 75% RH which are given as follows:

Tablet No.	1	2	3	4	5	6	7	8	9	10	11
Initial Weight of tablet	181	172	16	191	167	161	178	160	149	119	156
Weight of tablet after one month	211	210	210	203	196	190	191	177	173	170	163

Test whether weight of tablet is affected by 1 month storage at 45°C and 75% RH.

Solution:

Here parameter of interest is weight i.e. μ is the difference in weight at two time points (i.e., at the start of the experiment and after one month). We want to test if weight changes after one month storage at 45°C and 75% RH, so the null and alternate hypotheses will be

$$H_0: \mu = 0 \qquad \text{vs.} \qquad H_1: \mu \neq 0$$

Following table gives the initial weight, weight after one month of tablets and the difference between the initial weight and weight after one month.

Table: Required calculations for paired t test

Tablet No.	Initial Weight of tablet (milligrams) (x_{1i})	Weight of tablet after one month (milligrams) (x_{2i})	$d_i (x_{1i} - x_{2i})$
1	181	211	30
2	172	210	38
3	16	210	14
4	191	203	12
5	167	196	29
6	161	190	29
7	178	191	13
8	160	177	17
9	149	173	24
10	119	170	51
11	156	163	07
		Mean ($\bar{d}$)	24
		sd (s)	13.09

Here,

Mean = $\bar{d}$ = 24, sd = 13.09 and n = 11. Now

$$t = \frac{\bar{d}}{s/\sqrt{n}} = \frac{24}{13.09/\sqrt{11}} = 6.08$$

At 5% level of significance (α) = 0.05 (two sided) and degrees of freedom = 11-1 = 10, t = 2.228 (From "t" distribution tables).

As t(calculated) (=6.08) > t(tabulated) (=2.228), we reject the null hypothesis that there is no difference between the initial weight of rifampicin tablet and weight of rifampicin tablet after storing at 45°C and 75% RH for one month. This means that rifampicin tablets are not stable at 45°C and 75% RH if stored for one month.

Before performing the test, inspection of the data reveals the after storage weight values to be higher than the initial values for each of 11 tablets i.e. all the 11 differences are positive and a good portion are of substantial magnitude. On this basis, one would anticipate that the test of significance would yield a statistically significant result. The test confirms this expectation.

Example: Weight changes of 12 rats after being subjected to a regime of forced exercise are given below:
1.7, 0.7, -0.4, -1.8, 0.2, 0.9, -1.2, -0.9, -1.8, -1.4, -1.8 and -2.0 gms.
Each weight change is the weight after exercise minus the weigh before exercise. Test whether exercise reduces the weight or not.

Solution:
Here parameter of interest is weight i.e. μ is the difference in weight at two time points (i.e., after and before exercise regimen). We want to test if weight changes after exercise, so the hypotheses will be

$$H_0: \mu = 0 \qquad \text{vs.} \qquad H_1: \mu \neq 0$$

Here,

Mean = $\bar{d}$ = -0.65, sd = 1.57 and n = 12. Now

$$t = \frac{\bar{d} - \Delta}{s/\sqrt{n}} = \frac{-0.65 - 0}{1.5682/\sqrt{12}} = -1.81$$

At 5% level of significance (α) and degrees of freedom = 12-1 = 11, -t(tabulated) = -2.201 from "t" distribution tables.

As t(calculated) (=-1.81) > -t(tabulated) (=-2.201), we accept the null hypothesis that there is no difference in weight before and after exercise, i.e., the exercise regime does not reduce the weight based on this data.

IMPORTANCE OF CHOOSING THE CORRECT HYPOTHESIS

To understand the importance of correctly defining the problem, let us consider the following example:

Example: For a routine in house testing of weight of CBZ tablets, where the target weight of a batch is 325 mg, with a known population sd = 10.56, a sample of 20 tablets was collected at random and each tablet was weighed. Weights of 20 randomly selected tablets are
300, 316, 321, 325, 306, 317, 335, 325, 310, 319, 322, 327, 315, 320, 323, 331, 316, 320, 325 and 336
Find out whether the target weight is achieved based on this data.

Solution: Let μ be the average weight of tablets. Two objectives are possible:

 (i) Weight of tablet is equal to 325 mg or not, i.e., null and alternative hypotheses will be
    $H_0: \mu = 325$ vs. $H_1: \mu \neq 325$

 (ii) Weight of tablet should not exceed 325 mg, i.e., null and alternative hypotheses will be
    $H_0: \mu \geq 325$ vs $H_1: \mu < 325$

Here n = 20

$$\text{Mean} = \bar{x} = \frac{\sum_{i=1}^{n} x_i}{n} = (300+306+\ldots\ldots+336)/20 = 320.5 \text{ mg}$$

For the case (i)

$$Z = \frac{\bar{x} - \mu_0}{\sigma/\sqrt{n}} = \frac{320.5 - 325}{10.56/\sqrt{20}} = -1.906$$

$Z_{\alpha(2)}$ is 1.96 at 5% level of significance from "Z" distribution tables. As calculated Z statitic is -1.906 which is greater than -1.96, the null hypothesis will be accepted.

For the case (ii)

$$Z = \frac{\bar{x} - \mu_0}{\sigma/\sqrt{n}} = \frac{320.5 - 325}{10.56/\sqrt{20}} = -1.906$$

$-Z_{\alpha(1)}$ is -1.65 at 5% level of significance. As calculated Z statitic is -1.906 which is less than -1.65, the null hypothesis will be rejected.

This example highlights that hypotheses should be correctly defined as per the objectives, otherwise we may end up concluding incorrectly.

IMPORTANCE OF CHOOSING THE CORRECT STATISTICAL TEST

Example: In animals breeding experiment, the target weight of rats at a particular age was 325 g. Twenty (20) animals were sampled and mean weight of the animals was found to be 320.5 g. Objective was to test whether or not the mean weight of animals is equal to 325 g. Assume standard deviation as 9.95.

Solution: Let μ be the average weight of animals. The null hypothesis and alternate hypotheses are

$$H_0: \mu = 325 \text{ vs. } H_1: \mu \neq 325$$

(i) When 9.95 is the population sd
In case of known population sd, Z test should be used. Hence

$$Z = \frac{\bar{x} - \mu_0}{\sigma/\sqrt{n}} = \frac{320.5 - 325}{9.95/\sqrt{20}} = -2.02257$$

$Z_{\alpha/2}$ is 1.96 at 5% level of significance from "Z" distribution tables. As Z(calculated) is $< -Z_{\alpha/2}$, the null hypothesis will be rejected.

(ii) When 9.95 is the sample sd
In case of unknown population sd, t test should be used based on sample sd. Hence

$$t = \frac{\bar{x} - \mu_0}{\sigma/\sqrt{n}} = \frac{320.5 - 325}{9.95/\sqrt{20}} = -2.02257$$

$t_{\alpha(2),19}$ is 2.093 at 5% level of significance from "t" distribution tables. As t(calculated) is $> -t_{\alpha/2}$, the null hypothesis will be accepted.

Here we can see that if Z test is applied, it concludes that animal weight is not equal to 325 g, whereas when t test is applied it results in declaring that animal weight is 325 g.

So this example underlines the importance of choosing correct statistical tests.

To summarize

- Choosing the right hypothesis is very important as it can change the results of a test
- Be careful when apply these statistical tests as all the tests are based on some assumptions
  - When population variance(s) is known, Z tests are applied
  - When population variance(s) is not known, t tests are applied
- The statistical test chosen is dependent on the type of question being asked

What do you call a tea party with more than 30 people?

⋮

⋮

⋮

A "Z party"!!!

It is incorrect to say "Lie, damn lie and statistics" as problem is not with statistics but with misuse of it.

— Anonymous

Chapter 9

Multi Sample Hypotheses

Suppose an experiment consists of more than two treatments to be compared i.e. we have data from k (>2) samples. Then what should be the testing procedure? A very common or obvious answer to this is to apply two sample test multiple times. At a glance it seems OK.

Let us see what happens if we apply two sample test multiple times. Suppose there are following observations from k samples:

	$x_{11}, x_{12}, ..., x_{1n_1}$	$x_{21}, x_{22}, ..., x_{2n_2}$		$x_{(k-1)1}, x_{(k-1)2}, ..., x_{(k-1)n_k}$	$x_{k1}, x_{k2}, ..., x_{kn_k}$
Mean	$\bar{x}_1$	$\bar{x}_2$	..	$\bar{x}_{k-1}$	$\bar{x}_k$
Sample size	n_1	n_2	..	n_{k-1}	n_k
Sum of Squares	ss_1	ss_2	..	ss_{k-1}	ss_k
Variance	s_1^2	s_2^2	..	s_{k-1}^2	s_k^2

Suppose interest is in testing the means of various samples at level of significance 0.05, i.e.,

$$H_0: \mu_1 = \mu_2 = ... = \mu_{k-1} = \mu_k$$
vs.

H_1: At least one pair of means is not equal

As level of significance is 0.05, this implies

Probability [H_0 will be rejected when H_0 is true] = level of significance = 0.05

So the probability of accepting H_0 when H_0 is true will be

Probability [H_0 will be accepted when H_0 is true]

= 1 - Probability [H_0 will be rejected when H_0 is true]
= 1-.05
= 0.95

Suppose there are only 3 groups (G1, G2 and G3) to compare, that means 2-sample t-tests have to be applied 3 times for 3 different pairs of groups (test1: G1 & G2, test2: G2 & G3 and test3: G3 & G1). In this case, probability of accepting H_0 when H_0 is true will be the product of probabilities of accepting H_0 when H_0 is true for all the three tests, i.e.,

Probability [H_0 will be accepted when H_0 is true]
=Probability$_{test\ 1}$[H_0 will be accepted when H_0 is true] x Probability$_{test\ 2}$[H_0 will be accepted when H_0 is true] x Probability$_{test\ 3}$[H_0 will be accepted when H_0 is true] =0.95*0.95*0.95 =0.86

Probability [at least one test will reject H_0 when H_0 is true]
= 1- Probability [H_0 will be accepted in all the 3 tests when H_0 is true]
= 1 - 0.86
= 0.14

This implies that if 2-sample tests are applied 3 times for 3 different pairs of groups, the probability of rejecting at least one null hypothesis when null hypothesis is true (i.e. probability of committing at least one type one error) becomes 0.14 instead of 0.05.

$$\Downarrow$$

There is type I error inflation if 2-sample t-tests are applied multiple times to compare groups taking two groups at a time.

Following table gives the probability of committing at least one type one error when number of groups to be compared is k = 2, 3, 4, 5, 6, 10 and number of times, two sample t tests to be applied as K = 1, 3, 6, 10, 15 and 45.

Table: Probability of committing at least one type one error when apply 2 sample t-test multiple times where k = number of groups to be compared and K= number of times, 2 sample t tests to be applied

k	K	Level of significance				
		0.1	0.05	0.01	0.005	0.001
2	1	0.10	0.05	0.01	0.005	0.001
3	3	0.27	0.13	0.03	0.015	0.003
4	6	0.47	0.26	0.06	0.030	0.006
5	10	0.65	0.40	0.10	0.049	0.010
6	15	0.79	0.54	0.14	0.072	0.015
10	45	0.99	0.90	0.36	0.202	0.044
	∞	1.00	1.00	1.00	1.00	1.00

To avoid this "Error Inflation" in the case of multiple sample hypotheses, analysis of variance (ANOVA) is recommended as ANOVA protects the researcher against "Error Inflation".

ANALYSIS OF VARIANCE (ANOVA)

It is a technique
- Whereby the total variation present in a data set is partitioned or segregated into several components.
- Usually each of these components of variation is associated with a specific source of variation. Ratio of magnitude of variation due to each of these sources to the total variation in the data set is the basis of ANOVA.

For example, if we are looking for efficacy of a new drug and the trial is conducted in males and females of various age, ANOVA will segregate the variations in response of drug due to drug, age, sex etc.

Similar to other tests, ANOVA can be applied under few assumptions:
- All the samples are from normal distributions
- Observations are independent
- Variances of all the groups are same

Here assumption of normal distribution and independence can be verified by looking at the parameter of interest, experience or literature information. But equality of variances has to be tested using Bartlett's test as follows:

Bartlett's test for equality of variances

Let k samples are from k populations with variances $\sigma_1^2, \sigma_2^2, \sigma_3^2, \ldots\ldots\sigma_{k-1}^2, \sigma_k^2$. Then the null hypothesis and alternative hypothesis are

$$H_0 : \sigma_1^2 = \sigma_2^2 = \sigma_3^2 = \ldots\ldots = \sigma_{k-1}^2 = \sigma_k^2$$

vs.

$$H_1 : \sigma_i^2 \neq \sigma_j^2 \qquad \text{for any combination of i and j where } i \neq j.$$

Testing procedure:

- The test statistic for Bartlett's test is given by:

$$B = \frac{2.30256 * [\log s_p^2 * \sum\limits_{i=1}^{k}(n_i - 1) - \sum\limits_{i=1}^{k}(n_i - 1) * \log s_i^2]}{1 + [1/\{3*(k-1)\}] * [\sum\limits_{i=1}^{k}\frac{1}{(n_i - 1)} - \frac{1}{\sum\limits_{i=1}^{k}n_i}]}$$

where $s_p^2 = \sum_{i=1}^{k} ss_i \Big/ \sum_{i=1}^{k} (n_i - 1)$

- The test statistic B follows a distribution known as χ^2 with (k-1) df
- Null hypothesis of equality of variances will be rejected if B is greater than χ^2 (From "χ^2" distribution tables) at (k-1) degrees of freedom and α level of significance.
- If null hypothesis of equality of variances is accepted,
    o Parametric ANOVA will be applied
    o Otherwise non parametric ANOVA will be applied

ONE WAY ANOVA

One way ANOVA is applicable when there is only one factor which affects the response. For example, if it is known that effect of a particular drug is not dependent on age, gender or disease factor and we want to compare the various dose levels of that drug, we look only for the factor dose levels.

The ANOVA methodology is shown below through examples:

Example: An experiment was conducted to test the effect of 4 dose levels of a drug on kidney weight of dogs. Four (4) groups of dogs were taken and one dose level was administered to each group. After the assigned period, kidney weights were taken as follows. Kidney weights (X) are expressed as percentage of body weight.

Dose levels			
400 ppm	**200 ppm**	**100 ppm**	**0 ppm**
0.43	0.49	0.34	0.34
0.52	0.48	0.40	0.32
0.43	0.40	0.42	0.33
0.55	0.34	0.40	0.39

Test whether the 4 dose levels have same effect on kidney weight of dogs.

Solution:
Let μ_i for i=1, 2, 3, 4, are the mean kidney weight of 4 groups after giving the 4 dose levels. As per the objective, null and alternative hypotheses will be

$$H_0: \mu_1 = \mu_2 = \mu_3 = \mu_4$$
vs.
$$H_1: \text{At least one pair of means is not equal}$$

Before testing for equality of dose level effect, let us first test whether variance are equal or not using Bartlett's test, i.e.,

$$H_0 : \sigma_1^2 = \sigma_2^2 = \sigma_3^2 = \sigma_4^2$$

$$vs.$$

$$H_1 : \sigma_i^2 \neq \sigma_j^2 \qquad \text{for any combination of i and j where } i \neq j.$$

Table: Required calculations for Bartlett's test

Dose level	Mean kidney weight	s_i^2	ss_i	n_i	$n_i - 1$	$\log(s_i^2)$	$(n_i - 1) * \log(s_i^2)$	$1/(n_i - 1)$
400 ppm	0.4825	0.00383	0.0115	4	3	-2.4174	-7.2521	0.33333
200 ppm	0.4275	0.00502	0.0151	4	3	-2.2989	-6.8966	0.33333
100 ppm	0.3900	0.00120	0.0036	4	3	-2.9208	-8.7625	0.33333
0 ppm	0.3450	0.00097	0.0029	4	3	-3.0147	-9.0442	0.33333
Total	1.6450	0.01102	0.0331	16	12	-10.6518	-31.9553	1.33332

$$s_p^2 = \sum_{i=1}^{k} ss_i \Big/ \sum_{i=1}^{k} (n_i - 1) = 0.03305/12 = 0.00275$$

$$B = \frac{2.30256 * [\{\sum (n_i - 1)\} * \log s_p^2 - \sum (n_i - 1) * \log s_i^2]}{1 + [1/\{3 * (k-1)\}] * [\sum \dfrac{1}{(n_i - 1)} - \dfrac{1}{\sum n_i}]}$$

$$B = \frac{2.30256 * [12 * (-2.56) - (-31.955)]}{1 + [1/\{3 * (4-1)\}] * [1.3333 - 1/16]} = 2.4922$$

χ^2 is 7.815 at 5% level of significance and 3 degrees of freedom (From "χ^2" distribution tables). As calculated statistic B is less than tabulated χ^2, accept the null hypothesis of equal variance.

So we will perform parametric ANOVA calculations as follows:

$$\text{Correction factor (CF)} = \frac{(\sum_{i=1}^{k} \sum_{j=1}^{n_i} x_{ij})^2}{\sum_{i=1}^{k} n_i} = \frac{(6.58)^2}{4+4+4+4} = 2.7060$$

$$\text{Total sum of squares} = \text{SS(Total)} = \sum_{i=1}^{k} \sum_{j=1}^{n_i} x_{ij}^2 - CF = 2.7798 - 2.7060 = 0.0738$$

Sum of squares due to groups = SS(Between groups) = SS_{bg}

$$= \sum_{i=1}^{k} \frac{\left(\sum_{j=1}^{n_i} x_{ij}\right)^2}{n_i} - CF = \frac{1.93^2}{4} + \frac{1.71^2}{4} + \frac{1.56^2}{4} + \frac{1.38^2}{4} - 2.7060 = 0.04075$$

Sum of squares due to error = SS(Within groups) = SS_{wg}

$$= \text{SS(Total)} - \text{SS(Between groups)}$$
$$= 0.07380 - 0.04075$$
$$= 0.03305$$

Degree of freedom (Total) = df_{Total}
$$= n_1 + n_2 + n_3 + n_4 - 1$$
$$= 4 + 4 + 4 + 4 - 1$$
$$= 15$$

Degree of freedom (between groups) = df_{bg} = $k - 1 = 4 - 1 = 3$

Degree of freedom (within groups or error) = df_{wg} = $df_{Total} - df_{bg} = 15 - 3 = 12$

Mean sum of squares due to groups = $MS_{bw} = \dfrac{SS_{bg}}{df_{bg}} = \dfrac{0.04075}{3} = 0.01358$

Mean sum of squares due to error (MSE) = $MSwg = \dfrac{SS_{wg}}{df_{wg}} = \dfrac{0.03305}{12} = 0.00275$

$$F = \frac{MS_{bg}}{MS_{wg}} = \frac{0.01358}{0.00275} = 4.94$$

On summarizing all the calculations in Analysis of Variance table as follows:

Table: ANOVA table

Source of variation	df	SS	MS	F_{cal}	$F_{0.05(1),3,12}$
Between drugs	3	0.04073	0.01358	4.94	3.49
Error	12	0.03305	0.00275		
Total	15	0.07378			

As calculated F value is greater than tabulated F value, we will reject the null hypothesis of equality of mean kidney weight for all the dose levels. This implies that different dose levels have different effect on kidney weight.

Example: During method development for aceclofenac tablet, 100 mg (which was not official in any pharmacopoeia) aceclofenac was assayed by three different methods to find suitability of various methods. For each method, 5 tablets were assayed, results of which are as follows:

S.No.	Method A	Method B	Method C
1	102	99	103
2	101	100	100
3	101	99	99
4	100	101	104
5	102	98	102

Does these assay methods give different results, so as to decide whether any of these method should be dropped out?

Solution: Let μ_i for i=1, 2, 3, are the mean assay of 3 methods. For the problem in hand, Null hypothesis and alternate hypothesis are

H_0: $\mu_1 = \mu_2 = \mu_3$

H_1: Mean assay results are different for any two methods

Before testing for equality of effect, let us first test whether variance are equal or not using Bartlett's test, i.e.,

$$H_0 : \sigma_1^2 = \sigma_2^2 = \sigma_3^2$$

$$vs.$$

$$H_1 : \sigma_i^2 \neq \sigma_j^2 \qquad \text{for any combination of i and j where } i \neq j.$$

Table: Required calculations for Bartlett's test

Dose level	Mean	s_i^2	ss_i	n_i	$n_i - 1$	$\log(s_i^2)$	$(n_i - 1)*\log(s_i^2)$	$1/(n_i - 1)$
Method A	0.8367	0.7	2.8	5	4	-0.1549	-0.6196	0.25
Method B	1.1402	1.3	5.2	5	4	0.1139	0.4558	0.25
Method C	2.0736	4.3	17.2	5	4	0.6335	2.5339	0.25
Total	4.0505	6.3	25.2	15	12	0.5925	2.3700	0.75

$$s_p^2 = \sum_{i=1}^{k} ss_i \bigg/ \sum_{i=1}^{k} (n_i - 1) = 25.2/12 = 2.1$$

$$B = \frac{2.30256 * [\{\sum (n_i - 1)\} * \log s_p^2 - \sum (n_i - 1) * \log s_i^2]}{1 + [1/\{3 * (k-1)\}] * [\sum \frac{1}{(n_i - 1)} - \frac{1}{\sum n_i}]}$$

$$B = \frac{2.30256 * [12 * (0.3222) - (2.370)]}{1 + [1/\{3 * (3-1)\}] * [0.75 - 1/15]} = 3.093$$

χ^2 is 5.991 at 5% level of significance and 2 degrees of freedom (From "χ^2" distribution tables). As calculated statistic B is less than tabulated χ^2, accept the null hypothesis of equal variance.

Hence we will perform parametric ANOVA calculations as follows:

Table: Required calculations for analysis of variance

Sr. No.	Method A	Method B	Method C
1	102	99	103
2	101	100	100
3	101	99	99
4	100	101	104
5	102	98	102
Mean	101.2	99.4	101.6
sd	0.84	1.14	2.07
Total	506	497	508
Grand Total	1511		

$$\text{Correction factor (CF)} = \frac{(\sum_{i=1}^{k} \sum_{j=1}^{n_i} x_{ij})^2}{\sum_{i=1}^{k} n_i} = \frac{(1511)^2}{5 + 5 + 5} = 152208.1$$

$$\text{Total sum of squares} = SS(Total) = \sum_{i=1}^{k} \sum_{j=1}^{n_i} x_{ij}^2 - CF$$

$$= (102^2 + 101^2 + 101^2 + 100^2 + ...+ 100^2 + 102^2) - 152208.07$$
$$= 152247 - 152208.07$$
$$= 38.93$$

Sum of squares due to groups = SS(Between groups) = SS_{bg}

$$= \sum_{i=1}^{k} \frac{(\sum_{j=1}^{n_i} x_{ij})^2}{n_i} - CF = \frac{506^2}{5} + \frac{497^2}{5} + \frac{508^2}{5} - 152208.07 = 13.73$$

Sum of squares due to error = SS(Within groups) = SS_{wg}

$$= SS_{Total} - SS(Between\ groups)$$
$$= 38.93 - 13.73$$
$$= 25.20$$

Degree of freedom (Total) = df_{Total}
$$= n_1 + n_2 + n_3 - 1$$
$$= 5 + 5 + 5 - 1$$
$$= 14$$

Degree of freedom (between groups)= df_{bg} = k – 1 = 3 –1 = 2

Degree of freedom (within groups or error) = df_{WG} = df_{Total} - df_{bg} = 14 - 2 = 12

Mean sum of squares due to groups = MS_{bw} = $\dfrac{SS_{bg}}{df_{bg}}$ = $\dfrac{13.73}{2}$ = 6.87

Mean sum of squares due to error (MSE) = MS_{wg} = $\dfrac{SS_{wg}}{df_{wg}}$ = $\dfrac{25.20}{12}$ = 2.10

$$F = \frac{MS_{bg}}{MS_{wg}} = \frac{6.87}{2.10} = 3.27$$

Analysis of variance table will look like as follows:

Table: ANOVA table

Source of variation	df	SS	MS	F_{cal}	$F_{0.05(1),2,12}$
Between methods	2	13.73	6.87	3.27	3.89
Within methods (Error)	12	25.20	2.10		
Total	14	3 8.93			

As F(calculated) is less than F(tabulated), null hypothesis is accepted. That means all the three methods give same results. Hence any of these methods can be used for aceclofenac assay.

TWO WAY ANOVA

Two way ANOVA is applicable when there are two factors which affect the response. For example if it is known that effect of a particular drug is dependent on gender and we want to compare the various dose levels of that drug, we look for two factors as dose levels and gender.

The calculations are on the similar lines. Only we will have more terms as follows:

Let there be two factors A and B with levels a and b, respectively, and each group has equal number of experimental units n. The formulae will be as follows:

Correction factor (CF)

$$= \frac{(\sum_{i=1}^{a}\sum_{j=1}^{b}\sum_{l=1}^{n} x_{ijl})^2}{\sum_{i=1}^{a}\sum_{j=1}^{b} n_{ij}} \quad \text{where } n_{ij} = n \text{ for all } i = 1,\ldots., a \text{ and } j = 1,\ldots.,b$$

$$= \frac{(\sum_{i=1}^{a}\sum_{j=1}^{b}\sum_{l=1}^{n} x_{ijl})^2}{abn}$$

$$SS_{Total} = \sum_{i=1}^{a}\sum_{j=1}^{b}\sum_{l=1}^{n} x_{ijl}^2 - CF$$

$$SS(\text{Between groups of factor A}) = \sum_{i=1}^{a} \frac{(\sum_{j=1}^{b}\sum_{l=1}^{n} x_{ijl})^2}{bn} - CF$$

$$SS(\text{Between groups of factor B}) = \sum_{j=1}^{b} \frac{(\sum_{i=1}^{a}\sum_{l=1}^{n} x_{ijl})^2}{an} - CF$$

$$SS(\text{Cells}) = \frac{\sum_{i=1}^{a}\sum_{j=1}^{b}(\sum_{l=1}^{n} x_{ijl})^2}{n} - CF$$

$$SS(\text{Within cells}) = SS_{Total} - SS(\text{Cells})$$

SS(AxB interaction)
= SS (Cells) - SS(Between groups of factor A) - SS(Between groups of factor B)

Degrees of freedom are also calculated in the similar way, i.e.,

$$DF\ (\text{Total}) = N-1, \text{ where } N = \sum_{i=1}^{a}\sum_{j=1}^{b} n_{ij} = abn$$

DF (Factor A) = a-1
DF (Factor B) = b-1
DF (Cells) = ab -1
DF (AxB interaction) = ab-1-a+1-b+1 = ab-a-b+1 = (a-1)(b-1)

DF (Within Cells) = abn − ab = ab(n-1)

We will not perform calculation for this case or higher order case as these are tedious. Many softwares are available for that. Even MS Office Excel can be used for the purpose.

MULTIPLE COMPARISON

When null hypothesis of equality of means is rejected, we would like to know which group or groups have different mean compared to others. So our next objective is to perform pair wise comparison.

There are various methods as follows:
- The Tukey's Test
- The Newman-Keuls Test
- The Dunnett's test

The Tukey's test
- It considers the null hypothesis of equality versus non-equality of any two means
- First rank the samples by means from smallest to largest
- Calculate standard error (SE) using MSE,

$$SE = \sqrt{\frac{MSE}{n}} \qquad \text{if k groups are of equal size}$$

$$SE = \sqrt{\frac{MSE}{2}\left(\frac{1}{n_A}+\frac{1}{n_B}\right)} \quad \text{if k groups are not of equal size, SE will be}$$

different for each combination of means

- Calculate q as $q = \dfrac{\overline{x}_B - \overline{x}_A}{SE}$ for each combination of means

- Look for the $q_{\alpha,v,k}$ value which is dependent upon α, error degrees of freedom v from ANOVA and number of means k to be compared from the tables available in standard books of statistics. If calculated q for any combination of means is more than the $q_{\alpha,v,k}$, reject the equality of that particular combination of means.

To understand the method, let us use the data of first example of one way ANOVA analysis.

Sample means	0.4825	0.4275	0.39	0.345
Samples ranked by means	4	3	2	1

$$SE = \sqrt{\frac{s^2}{n}} = \sqrt{\frac{0.00275}{4}} = 0.02622$$

Table: Illustration of Tukey's method

Comparison	Difference	SE	q	$q_{0.05,12,4}$	Conclusion
(B vs. A)	$\overline{x}_B - \overline{x}_A$				
4vs1	0.1375	0.02622	5.30	4.199	4 & 1 are statistically significantly different
4vs2	0.0925	0.02622	3.53	4.199	4 & 2 are not statistically significantly different
4vs3	0.1375	0.02622	2.10	4.199	4 & 3 are not statistically significantly different
3vs1	0.0820	0.02622	3.15	4.199	3 & 1 are not statistically significantly different
3vs2	0.0375	0.02622	1.43	4.199	3 & 2 are not statistically significantly different
2vs1	0.0450	0.02622	1.72	4.199	2 & 1 are not statistically significantly different

This analysis concludes that only group 4 and 1 are statistically significantly different.

If no difference is found between two means, it is concluded that there is no difference between any means enclosed by these two. As in the above example, no statistically significant difference was found between 4 and 2, so there is no need to test further for 4 vs 3. Similarly there is no need to test 3 vs 2.

The Newman-Keuls Test

This is same as Tukey's test except that the k in the critical value $q_{\alpha,v,k}$ is the number of means in the range of means tested. For example, if $\overline{X}_5$ is compared with $\overline{X}_1$, the k will be 5; if $\overline{X}_5$ is compared with $\overline{X}_2$, the k will be 4.

For example, in the above example, all the calculations are same except the critical values as given below.

Table: Illustration of Newman-Keuls method

Comparison	Difference	SE	q	k	$q_{0.05,12,k}$	Conclusion
(B vs. A)	$\overline{x}_B - \overline{x}_A$					
4vs1	0.1375	0.02622	5.30	4	4.199	4 & 1 are statistically significantly different

4vs2	0.0925	0.02622	3.53	3	3.773	4 & 2 are not statistically significantly different
4vs3	0.1375	0.02622	2.10	2	3.082	4 & 3 are not statistically significantly different
3vs1	0.0820	0.02622	3.15	3	3.773	3 & 1 are not statistically significantly different
3vs2	0.0375	0.02622	1.43	2	3.082	3 & 2 are not statistically significantly different
2vs1	0.0450	0.02622	1.72	2	3.082	2 & 1 are not statistically significantly different

In this example, both the methods i.e., Tukey's test and Newman-Keuls Test give the same results. But this is not always true. Different researchers have different opinion about these tests i.e. some prefer Tukey's test whereas others prefer Newman-Keuls Test.

The Dunnett's test

When our interest is in comparing all the means with one mean known as "control", this method is applied. If there are k groups to be compared, there will be (k-1) comparisons.

- Calculate standard error (SE) using MSE,

$$SE = \sqrt{\frac{2*MSE}{n}}$$ if k groups are of equal size

$$SE = \sqrt{MSE(\frac{1}{n_A} + \frac{1}{n_{control}})}$$ if k groups are not of equal size, SE will be different for each mean

- Calculate q as $q = \dfrac{\overline{X}_{control} - \overline{X}_A}{SE}$ for each mean

- Look for the $q'_{\alpha,v,k}$ value which is dependent upon α, error degrees of freedom v from ANOVA and number of means k to be compared from the tables available in standard books of statistics. If calculated q for a combination of mean and control is more than the $q'_{\alpha,v,k}$, reject the null hypothesis of equality of that mean with control.

ANALYSIS OF COVARIANCE (ANCOVA)

Analysis of covariance is performed when it is known that there is a variable(s) which affects the outcome indirectly. Examples,

- Initial weight may have effect on two diet programs for losing weight

- Race may play a role in comparison of 16 year old males' blood pressure of two countries

In these cases, analysis is performed taking these variables (initial weight, race) as covariate. However, we will not discuss the method of analysis as it can be performed using any software.

To summarize

- If number of groups is more than two, use ANOVA
- Test for equality of variances among groups before applying ANOVA
- In case of significant results in ANOVA, go for multiple comparison

Statistics means never having to say you're certain.

— **Anonymous**

Chapter 10

Non-Parametric Tests

- When assumption of normal distribution of the data is not satisfied, non parametric tests are used. these are known as non-parametric tests because these do not require any assumption regarding the distribution (distributions are defined by parameters) of the data. Example, clinical evaluation parameters graded as 0, 1, 2 and 3.
- As these tests do not require the data having any specific distribution, non-parametric tests also sometimes, known as distribution free methods.
- Similar to parametric tests, there are different-2 non-parametric tests for various situations as follows:

SINGLE SAMPLE TEST: WILCOXON SIGNED RANK TEST

This test is applicable for single sample case (it is similar to one sample case of parametric test). The null and alternative hypotheses can be:

(i) H_0: Median of the group is μ_0 vs. H_1: Median of the group is not μ_0

(ii) H_0: Median of the group $\geq \mu_0$ vs. H_1: Median of the group $< \mu_0$

(iii) H_0: Median of the group $\leq \mu_0$ vs. H_1: Median of the group $> \mu_0$

where μ_0 is some hypothesized median.

Testing procedure:
- Arrange the observations from smallest to the largest
- Rank them
    - o   "Tied observations are assigned an average rank"
- Assign a plus sign to each associated with a datum $> \mu_0$ or minus sign to each associated with a datum $< \mu_0$
- Determine the sum of the positive ranks (T+) and sum of the negative ranks (T-)

- Ignoring the sign, find smaller sum (T) of the two
- Reject the null hypothesis
  - if T is less than $T_{\alpha(2),n}$, the tabulated value, for case (i)
  - if T+ is less than $T_{\alpha(1),n}$, the tabulated value, for case (ii)
  - if T- is less than $T_{\alpha(1),n}$, the tabulated value, for case (iii)

where $T_{\alpha(2),n}$ and $T_{\alpha(1),n}$ are from Wilcoxon T tests tables, usually given in standard books of statistics for one or two tailed tests and various values of n.

TWO SAMPLE CASE: MANN-WHITNEY TWO SAMPLE TEST

This test is applicable when two independent samples have to be compared (it is similar to two sample case of parametric test). The null and alternative hypotheses can be:

(i) H_0: Measuements in group 1 are same as measuements in group 2

vs. H_1: Measuements in group 1 are not same as measuements in group 2

(ii) H_0: Measuements in group 1 $\geq$ Measuements in group 2

vs. H_1: Measuements in group 1 $<$ Measuements in group 2

(iii) H_0: Measuements in group 1 $\leq$ Measuements in group 2

vs. H_1: Measuements in group 1 $>$ Measuements in group 2

Testing procedure:
- First combine observations of both the samples
- Arrange the combined observations from smallest to the largest
- Rank them
- "Tied observations are assigned an average rank"
- Determine the sum of ranks for group 1(R_1) and group 2 (R_2)
- Determine Mann-Whitney test statistic as follow:

$$U = n_1 n_2 + \frac{n_1(n_1+1)}{2} - R_1,$$

$$U' = n_1 n_2 - U$$

  where n_1 and n_2 are the sample size for the two samples
- Reject the null hypothesis
  - if either U or U' is greater than $U_{\alpha(2),n_1,n_2}$, the tabulated value, for case (i)
  - if U is greater than $U_{\alpha(1),n_1,n_2}$, the tabulated value, for case (ii)
  - if U' is greater than $U_{\alpha(1),n_1,n_2}$, the tabulated value, for case (iii)

where $U_{\alpha(2),n_1,n_2}$ and $U_{\alpha(1),n_1,n_2}$ are from Mann-Whitney U test tables, usually given in standard books of statistics for one or two tailed tests and various values of n_1 and n_2. Generally tables assume that $n_1 > n_2$, so sample with larger sample size may be assumed as sample 1 and other one as sample 2.

Example: Approximate times to death (in hours) of rats after dosing different doses A and B are as follows

1, 2, 6, 1, 5, 1, 3, 2 hrs for dose A

2, 4, 1, 8, 7, 5, 4, 4 hrs for dose B

Find if approximate times to death of rats after dosing dose A and dose B are same or not.

Solution:

- The hypotheses are H_0: Approximate times to death of rats after dosing different doses A and B are same vs. H_1: Approximate times to death of rats after dosing different doses A and B are not same
- First combine the data of both the groups.

  1, 2, 6, 1, 5, 1, 3, 2, 2, 4, 1, 8, 7, 5, 4, 4
- Arrange them in increasing order

  1, 1, 1, 1, 2, 2, 2, 3, 4, 4, 4, 5, 5, 6, 7, 8
- Then rank them as follows:

Dose type	Hours to death	Rank	Assigning average rank to tied observations
Dose A	1	1	2.5
Dose A	1	2	2.5
Dose A	1	3	2.5
Dose B	1	4	2.5
Dose A	2	5	6
Dose A	2	6	6
Dose B	2	7	6
Dose A	3	8	8
Dose B	4	9	9.5
Dose B	4	10	9.5
Dose A	5	11	11.5
Dose B	5	12	11.5
Dose A	6	13	13
Dose B	7	14	14.5
Dose B	7	15	14.5
Dose B	8	16	16

Sum of Ranks for dose A $= R_1 = 52.0$

Sum of Ranks for dose B $= R_2 = 84.0$

$$U = n_1 n_2 + \frac{n_1(n_1+1)}{2} - R_1 = 8*8 + \frac{8*(8+1)}{2} - 52 = 48$$

$$U' = n_1 n_2 - U = 8*8 - 48 = 16$$

Tabulated value of the statistic at 5% level of significance, i.e., $U_{\alpha(2),n_1,n_2} = 51$

As neither U nor U' is greater than 51, the two dose groups are not considered significantly different at 5% level of significance, i.e., approximate times to death of rats after dosing dose A and dose B are same.

Example: To compare the % drug released at 45 minutes of dissolution test from two different batches of rifampicin tablets, six (6) tablets were taken from rifampicin batch A and rifampicin batch B. Observations are as follows:

% drug release from batch A at 45 minutes	% drug release from batch B at 45 minutes
62	67
60	68
61	65
65	66
69	67
68	69

Objective of the experiment is to test whether % drug release at 45 minutes from batches A and B is same.

Solution:

- The hypotheses are H_0: % drug release at 45 minutes from batches A and B are same vs. H_1: % drug release at 45 minutes from batches A and B are not same
- Arranging the observations from smallest to largest after combining them, 60, 61, 62, 65, 65, 66, 67, 67, 68, 68, 69, 69
- Assigning rank to them (observation-rank)
  60- 1, 61- 2, 62- 3, 65-4.5, 65-4.5, 66- 6, 67- 7.5, 67- 7.5, 68- 9.5, 68- 9.5, 69- 11.5, 69- 11.5.

Table: Determination of sum of rank

% drug release from A	Rank	% drug release from B	Rank
62	3	67	7.5
60	1	68	9.5

61	2	65	4.5
65	4.5	66	6
69	11.5	67	7.5
68	9.5	69	11.5
Total	31.5		46.5

Sum of ranks for batch A = R_1 = 31.5
Sum of ranks for batch B = R_2 = 46.5

$$U = n_1 n_2 + \frac{n_1(n_1+1)}{2} - R_1 = 6*6 + \frac{6*(6+1)}{2} - 31.5 = 25.5$$

$$U' = n_1 n_2 - U = 6*6 - 25.5 = 10.5$$

Tabulated value of the statistic at 5% level of significance, i.e., $U_{\alpha(2),n_1,n_2} = 31$

As neither U nor U' is greater than 31, the two dose groups are not considered significantly different at 5% level of significance, i.e., % drug release at 45 minutes from batches A and B is same.

PAIRED CASE: WILCOXON SIGNED RANK TEST

This test is applicable for paired samples case, i.e., when observations are taken on one object either on two occasions or from two sites (it is similar to paired sample case of parametric test).

Testing procedure:
- Calculate the difference for each pair of observations
  Once the difference is taken this becomes a single sample case. Hence the null and alternative hypotheses can be:
  (i) H_0: Median difference for each pair of observations is μ_0

   vs. H_1: Median difference for each pair of observations is not μ_0

  (ii) H_0: Median difference for each pair of observations $\geq \mu_0$

   vs H_1: Median difference for each pair of observations $< \mu_0$

  (iii) H_0: Median difference for each pair of observations $\leq \mu_0$

   vs H_1: Median difference for each pair of observations $> \mu_0$

  where μ_0 can assume any value based on the acceptable difference between the two sets of observations.
- Ignoring the signs of the differences, arrange the observations from smallest to the largest
- Rank them
  - "Tied observations are assigned an average rank"
  - "If there is a difference of 0, it is eliminated from the analysis and the sample size is reduced by 1 for each pair eliminated"

- Assign each rank a plus or minus sign depending on the sign of the difference
- Determine the sum of the positive ranks and sum of the negative ranks
- Ignoring the sign, find smaller sum (T) of the two
- Reject the null hypothesis
    - if T is less than $T_{\alpha(2),n}$, the tabulated value, for case (i)
    - if T+ is less than $T_{\alpha(1),n}$, the tabulated value, for case (ii)
    - if T- is less than $T_{\alpha(1),n}$, the tabulated value, for case (iii)

where $T_{\alpha(2),n}$ and $T_{\alpha(1),n}$ are from Wilcoxon T tests tables, usually given in standard books of statistics for one or two tailed tests and various values of n.

Example: In a bioequivalence study, same subject is given formulations to be compared on two occasions. Let there be two formulations A and B to be compared. Twelve (12) subjects were taken for the study and T_{max} was obtained for formulations A and B as follows. Test whether T_{max} for both the formulations is same or not.

Subject No.	1	2	3	4	5	6	7	8	9	10	11	12
T_{max} for A (hr)	2.5	3	1.25	1.75	3.5	2.5	1.5	2.25	3.5	2.5	2	3.5
T_{max} for B (hr)	3.5	4	2.5	2	3.5	4	1.5	2.5	3	3	3.5	4

Solution:
- The hypotheses are H_0: T_{max} for both the formulations are same vs. H_1: T_{max} for both the formulations are not same
- Calculate the difference for each pair of observations
- Arrange the observations from smallest to the largest ignoring the signs of the differences
- Rank them, assign each rank a plus or minus sign depending on the sign of the difference
- Determine the sum of the positive ranks and sum of the negative ranks

Table: Required calculations for Wilcoxon Signed rank test

Subject	A	B	B-A	Rank	Assigned rank	Assigned rank with sign
1	2.5	3.5	+1	7	7.5*	+7.5
2	3	4	+1	8	7.5*	+7.5
3	1.25	2.5	+1.25	9	9	+9
4	1.75	2	+.25	1	2	+2
5**	3.5	3.5	+0			
6	2.5	4	+1.5	10	10.5*	+10.5
7	1.5	1.5	-0.25	2	2	-2

8	2.25	2.5	+0.25	3	2	+2
9	3.5	3	-0.5	4	5	-5
10	2.5	3	+0.5	5	5	+5
11	2	3.5	+1.5	11	10.5	+10.5
12	3.5	4	+0.5	6	5	+5

*Tied observations are assigned average rank
**As difference is 0, it is eliminated from the analysis and sample size is reduced to 11

Sum of Positive ranks = 59
Sum of negative ranks =7
Ignoring the sign, smaller sum (T) of the two sums is 7
$T_{0.05(2),11} = 10$

Reject the null hypothesis as smaller sum (T=7) is less than tabulated value of T (=10). Hence T_{max} for both the formulations is not same.

Example: Hardness of Aceclofenac tablet from the same batch was measured using two different hardness testers i.e. Monsanto tester and Schleuniger tester. Hardness data was obtained for a total of 24 tablets, 12 tablets each of one hardness tester, which is as follows:

Sr. No	Hardness of tablets using Monsanto Hardness tester (A) (Newtons)	Hardness of tablets using Schleuniger Hardness tester (B) (Newtons)
1	2.5	3.50
2	3.0	4.00
3	1.25	2.50
4	1.75	2.00
5	3.50	3.50
6	2.50	4.00
7	1.75	1.50
8	2.25	2.50
9	3.50	3.00
10	2.50	3.00
11	2.00	3.50
12	3.50	4.00

Objective was to find out whether these hardness testers could be used interchangeably.

Solution:

- Based on the objectives, the hypotheses are H_0: Different hardness testers are same vs. H_1: Different hardness testers are not same
- Calculate the difference for each pair of observations
- Arrange, rank the observations from smallest to the largest ignoring the signs of the differences
- Assign each rank a plus or minus sign depending on the sign of the difference
- Determine the sum of the positive ranks and sum of the negative ranks

Table: Required calculations for Wilcoxon Signed rank test

Sr. No.	Hardness of tablets using Monsanto Hardness tester (A) (Newtons)	Hardness of tablets using Schleuniger Hardness tester (B) (Newtons)	Difference B-A	Rank	Ranks with (+) Signs	Ranks with (-) Signs
1	2.5	3.50	+ 1	7.5	7.5	
2	3.0	4.00	+1	7.5	7.5	
3	1.25	2.50	+1.25	9	9	
4	1.75	2.00	+0.25	2	2	
5	3.50	3.50	0			
6	2.50	4.00	+1.50	10.5	10.5	
7	1.75	1.50	-0.25	2		2
8	2.25	2.50	+0.25	2	2	
9	3.50	3.00	-0.50	5		5
10	2.50	3.00	+0.50	5	5	
11	2.00	3.50	+1.50	10.5	10.5	
12	3.50	4.00	+0.50	5	5	
				Total	59	7

Sum of positive ranks = 59

Sum of negative ranks = 7

$T_{0.05(2),11}$ (Tabulated) = 10

As minimum of the two rank sum i.e., 7 is less than 10, null hypothesis is rejected. Therefore, the difference is significant at 0.05 level of significance.

That means, both the hardness testers are significantly different and cannot be used interchangeably.

MULTI SAMPLE CASE: KRUSKAL - WALLIS NONPARAMETRIC ANOVA

As in parametric case, nonparametric ANOVA is performed when there are more than two groups to compare.

Testing procedure:
- The null and alternative hypotheses in this case are H_0: Measrements in all the groups are same vs. H_1: Measrements in all the groups are not same
- Arrange the combined observations from smallest to the largest
- Rank them
- Determine the sum of ranks for the each group (R_i, i=1,2,3,…,k)
- Reject the null hypothesis (that there is no statistically significant difference between the groups) if the statistic H given by

$$H = \frac{12}{N(N+1)} \sum_{i=1}^{k} \frac{R_i^2}{n_i} - 3(N+1)$$

is greater than the H obtained from the table for Kruskal Wallis test available in books for small sample sizes and $k \leq 5$. For $k > 5$, "H" may be considered to be approximated by χ^2 with k-1 degrees of freedom.

Example: An experiment was done to determine the effect of lubricant concentration on the hardness of the tablet and to set the specification limits for the lubricant in the formulation. Lubricant (Aerosil) concentration was varied at two levels i.e. low (0.5%) and high (1.5%), when the optimized % in the formulation was 1.00 %, in the formulation of Aceclofenac tablet, 100 mg. Hardness of 10 tablets (in Kp) was determined. In case of optimized batch, due to availability constrain, only 9 tablets were analyzed. The data obtained is as follows:

Sr. No.	Hardness (Kp) at 1 % w/w	Hardness (Kp) at 0.5 % w/w	Hardness (Kp) at 1.5 % w/w
1	8	10	3
2	1	5	4
3	9	8	8
4	9	6	1
5	6	7	1
6	3	7	3

7	15	15	1
8	1	1	6
9	7	15	2
10	-	7	2

Determine the effect of lubricant concentration on the hardness of the tablet.

Solution:

The null and alternative hypotheses in this case are H_0: Hardness of tablets is same for all lubricant concentrations vs. H_1: Hardness of tablets is not same for all lubricant concentrations

Table: Required calculations for Kruskal – Wallis ANOVA

S.No.	1 % w/w Control	Rank	0.5 % w/w	Rank	1.5 % w/w	Rank
1	8	22	10	26	3	10
2	1	3.5	5	13	4	12
3	9	24.5	8	22	8	22
4	9	24.5	6	15	1	3.5
5	6	15	7	18.5	1	3.5
6	3	10	7	18.5	3	10
7	15	28	15	28	1	3.5
8	1	3.5	1	3.5	6	15
9	7	18.5	15	28	2	7.5
10	-		7	18.5	2	7.5
Sum of Ranks		149.5		191.0		94.50

$$H = \frac{12}{N(N+1)} \sum_{i=1}^{k} \frac{R_i^2}{n_i} - 3(N+1)$$

$$= 12 / (29\,(29+1)) * (149.5^2/9 + 191^2/10 + 94.5^2/10) - 3(39+1)$$

$$= 6.89$$

As sample size is large, χ^2 is used for testing instead of H. χ^2 with 2 degrees of freedom at 5 % level of significance is 5.991 from the χ^2 tables. Therefore, null hypothesis of equality of hardness of tablets at different lubricant levels is rejected which implies that mean average hardness differs for at least two of the three treatment groups (1.00%, 1.5%, 0.5%) at 5% level of significance.

To summarize

Advantages of non-parametric tests
- Do not require the assumption that the underlying distribution is normal or any other form of distribution
- Can be performed quickly as they deal with ranks rather than actual values
- Use of ranks makes nonparametric techniques less sensitive to measurement error than traditional parametric tests
- Can be used with other than continuous type of data

Disadvantages of non-parametric tests
- If underlying assumptions are satisfied, nonparametric tests are less powerful than the comparable parametric tests

The most misleading assumptions are the ones you don't even know you're making.

— **Douglas Adams**

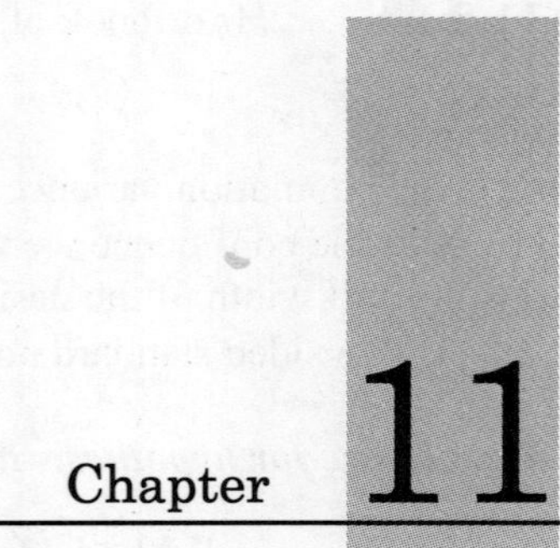

Chapter 11

Sample Size Estimation

A very common question raised by the experimentalist is what should be the sample size for a particular experiment. Number of units to be used in a scientific investigation is extremely important in pharmaceutical sciences because
(i) Using more number of experimental units involves more time, money and energy unnecessarily. Moreover, if it involves animals or humans, it is unethical.
(ii) Using less number of experimental units may not show a statistically significant difference due to less power. This results in wastage of all efforts put in the experiment.

Answer to this question is not straight forward as depending upon the design, depending upon the parameter type which are tested (i.e. whether the parameter is quantitative or qualitative; if quantitative discrete or continuous etc.), sample size is decided. However, we can define the critical factors influencing the sample size. Sample size depends on the following factors:
- Significance level (α)
- The β error or Power ($1-\beta$)
- The difference to be detected (Δ)
- Variability in the data (σ or %CV)

Sample size (n) estimation formulae for few situations are as follows:

SAMPLE SIZE FORMULAE FOR MEANS

One sample case with known population variance

Sample size for estimation of confidence interval for mean

$$n = \frac{\sigma^2 Z_{\alpha(2)}^2}{d^2}$$

where

σ^2 is the population variance

$(1 - \alpha)$ is the confidence we want in the estimate

d is the half width of the desired confidence interval

$Z_{\alpha(2)}$ is two sided standard normal variate at α level of significance

Sample size for hypothesis testing

$$n = (\frac{\sigma}{\delta})^2 (Z_\alpha + Z_{\beta(1)})^2$$

where

σ^2 is population variance

δ is the minimum difference to be detected

α can be either $\alpha(1)$ or $\alpha(2)$, respectively depending on whether one tailed or two tailed test is to be used

$(1 - \beta)$ is the desired power of the test

Z_α & $Z_{\beta(1)}$ are the values obtained from normal distribution tables at α and β levels.

One sample case with unknown population variance

Sample size for estimation of confidence interval for mean

$$n = \frac{s^2 t^2_{\alpha(2),(n-1)}}{d^2}$$

s^2 is the estimate of population variance σ^2

$(1 - \alpha)$ is the confidence we want in the estimate

$(n-1)$ is the degrees of freedom

d is the half width of the desired confidence interval

t is two sided t value from "t" distribution table at α level of significance and (n-1) degrees of freedom

Sample size for hypothesis testing

$$n = \frac{s^2}{\delta^2} (t_{\alpha,v} + t_{\beta(1),v})^2$$

where

s^2 is an estimate of population variance σ^2

δ is the minimum difference to be detected

α can be either $\alpha(1)$ or $\alpha(2)$, respectively depending on whether one tailed or two tailed test is to be used

$v = (n-1)$ is the degrees of freedom

$(1 - \beta)$ is the desired power of the test

$t_{\alpha,v}$ & $t_{\beta(1),v}$ are the values obtained from the t distribution table at α and β levels and v degrees of freedom

Two sample case with known population variance

Sample size for estimation of confidence interval for difference in two means

$$n = \frac{2\sigma^2 Z^2_{\alpha(2)}}{d^2}$$

where

n is the sample size for each group

σ^2 is the pooled population variance

$(1 - \alpha)$ is the confidence we want in the estimate

d is the half width of the desired confidence interval

$Z_{\alpha(2)}$ is two sided standard normal variate at α level of significance

Sample size for hypothesis testing between two means

$$n = 2(\frac{\sigma}{\delta})^2 (Z_\alpha + Z_{\beta(1)})^2$$

where

n is the sample size for each group

σ^2 is pooled population variance

δ is the minimum difference to be detected between two means

α can be either $\alpha(1)$ or $\alpha(2)$, respectively depending on whether one tailed or two tailed test is to be used

$(1 - \beta)$ is the desired power of the test

Z_α & Z_β are the values obtained from normal distribution tables at α and β levels.

Two sample case with unknown population variance

Sample size for estimation of confidence interval for difference in two means

$$n = \frac{2s_p^2 t^2_{\alpha(2), 2(n-1)}}{d^2}$$

where

n is the sample size for each group

s_p^2 is the pooled sample variance

$(1 - \alpha)$ is the confidence we want in the estimate

$v = 2(n-1)$ is the degrees of freedom

d is the half width of the desired confidence interval

$t_{\alpha(2), v}$ are the values obtained from the t distribution table at α level and v degrees of freedom

Sample size for testing between two means

$$n = 2(\frac{s_p}{\delta})^2 (t_{\alpha,v} + t_{\beta(1),v})^2$$

where
n is the sample size for each group

s_p^2 is pooled sample variance and an estimate of population variance σ^2

δ is the minimum difference to be detected between two means
α can be either $\alpha(1)$ or $\alpha(2)$, respectively, depending on whether one tailed or two tailed test is to be used
β is (1- power)
$v = 2(n-1)$ is the degrees of freedom
$t_{\alpha,v}$ & $t_{\beta(1),v}$ are the values obtained from the t distribution table at α and β levels and v degrees of freedom

Unequal sample sizes

Sometimes due to some constraint, size of one sample (n1) can not be increased, then we fix n_1 and n_2 is calculated as follows:

$$n_2 = \frac{nn_1}{2n_1 - n}$$

where n is sample size calculated from above mentioned formulae for equal sample size.

Example: Let an experiment was planned with 45 experimental units in two treatment groups ($n_1 = n_2 = 45$). But due to some reasons, only 30 experimental units can be taken in one group, then in another group number of subjects required will be

$$n_2 = \frac{45*30}{2*30-45} = \frac{1350}{15} = 90$$

Paired sample case

Sample size formulae for this case are same as for one sample case with σ^2 as the variance of difference in the paired observations rather than the variable itself.

Multi sample case (i.e. ANOVA)

Let here are k groups to be compared.
For this case, the calculation of sample size is iterative.
  - Guess a value of n
  - Calculate, $v_1= k-1$, $v_2= k(n-1)$ and

$$\varphi = \sqrt{\frac{n\delta^2}{2ks^2}}$$

where s^2 is the mean square error (MSE) of ANOVA, δ is minimum difference to be detected in means.

- Find the power corresponding to the value of φ, v_1 and v_2 from the figures for "power and sample size in analysis of variance"
- If the power is more than the required one, another guess for n will be lower
- otherwise higher value of n should be guessed
- Repeat steps 1-5 till we get the desired power

Example: In the experiment to determine whether the development time for insect embryos is the same at 3 different experimental temperatures, how many replicate data should be collected so as to have an 80% probability of detecting a difference between population means as 2 days, at 5% level of significance? We shall assume that $s^2=1.655$.

Solution:
- Here k=3, δ = 2 and $s^2=1.655$

- Let us guess n=15
    $\varphi=2.46$, $v_1= k-1= 3-1 = 2$, $v_2= k(n-1) =3*(15-1)=42$
    From the table, Power$\approx$0.96

- Next guess **n=10**
    $\varphi=2.01$, $v_1= k-1= 2$, $v_2= k(n-1) =3*(10-1)=27$
    Power$\approx$0.84

- Next guess **n=9**
    $\varphi=1.90$, $v_1= k-1= 2$, $v_2= k(n-1) =3*(9-1)=24$
    Power$\approx$0.79

Hence we conclude that using sample size of at least 10 should result in an ANOVA with a power of at least 0.84.

SAMPLE SIZE FORMULA FOR PROPORTION

Sample size for estimation of proportion
If p and q are not very close to 0, following formula is applicable

$$n = \frac{pqZ^2_{\alpha(2)}}{\delta^2}$$

where

        p is the proportion
        q = 1- p
        δ is the allowable error

$Z_{\alpha/2}$ is two sided standard normal variate at α level of significance

If estimate of p and q are not available, n can be estimated by

$$n = \frac{Z^2_{\alpha(2)}}{4\delta^2} \text{ on assuming p and q to be } 1/2.$$

Sample size for hypothesis testing between two proportions

$$n = \frac{[Z_\alpha \sqrt{2\overline{pq}} + Z_{\beta(1)} \sqrt{p_1 q_1 + p_2 q_2}]^2}{\delta^2}$$

where
p_1 and p_2 are sample proportions for the two samples
$q_1 = 1 - p_1$ and $q_2 = 1 - p_2$

$$\overline{p} = \frac{p_1 + p_2}{2}$$

$$\overline{q} = 1 - \overline{p}$$

δ is the difference to be detected between two proportions
α can be either $\alpha(1)$ or $\alpha(2)$, respectively, depending on whether one tailed or two tailed test is to be used
β is (1- power)
Z_α & Z_β are the values obtained from normal distribution tables at α and β levels.

As the above formula for sample size gives an under-estimate of n, following formula is recommended:

$$n' = \frac{n}{4}\left[1 + \sqrt{1 + \frac{4}{n\delta}}\right]^2$$

where
n is the sample size obtained from the above formula.

To summarize

- Always remember that the sample size formulae are different for different situations
- Be very clear where your problem fits, based on that, use the appropriate formula

The best thing about being a statistician is that you get to play in everyone's backyard.

— **John Wilder Tukey**

Chapter 12

Simple Linear Regression and Correlation

When two or more variables are measured either on the same subjects /animals / experimental units or on different subjects /animals / experimental units to see whether one is associated with another, it is described in terms of regression and correlation. In this chapter, we will consider the case of **two variables** only.

In pharmaceutical / medical sciences, it is frequently desirable to learn about the relationship between variables. For example,

- *In-vivo* and *in-vitro* data
- Drug concentration and absorbance
- Blood pressure and age
- Height and weight
- Dose administered and its response
- Concentration of drug in biological fluids and its response
- Linear portion of pharmacokinetic data
- Stability studies

HOW TO DETECT THE RELATIONSHIP

If changes in the values of one variable are matched with similar changes in another variable, there may be a relationship between the two variables.

Visual inspection: To detect if there is a correlation or relationship between two variables, the simplest approach is the visual inspection. We should plot the data having one variable on x axis and another variable on y axis (in case of two variables). If there appears a relationship between the variables, we measure the extent of relationship (correlation) and the form of relationship (regression).

However, always keep in mind that all the relationships are not causal i.e., correlation does not mean causation always. For example,

* Ice-cream sales in Britain and road accidents in France
* Annual sales of chewing gum and the incidence of crime in US

The above may have a correlation but there is no logic to be correlated.

This is spurious correlation

CORRELATION

Correlation is concerned with measuring the strength of the relationship between variables or degree of correlation between variables.

Characteristics of correlation:

* Correlation is of two types:
  * Positive: When change in both the variables is in one direction
  * Negative: When change in both the variables is in opposite direction
* Correlation lies between -1 and 1
* Correlation can be linear as well as non-linear, however, we will discuss only linear correlation

Linear correlation

If there is a linear relationship between two variables X and Y, it is measured by Pearson's Correlation Coefficient (r) defined as follows:

$$r = \frac{Cov(X, Y)}{\sqrt{Var(X)Var(Y)}}$$

$$= \frac{\sum_{i=1}^{n}(x_i - \bar{x})(y_i - \bar{y})}{\sqrt{\sum_{i=1}^{n}(x_i - \bar{x})^2 \sum_{i=1}^{n}(y_i - \bar{y})^2}}$$

(n is the number of observations on X and Y)

$$= \frac{\sum x_i y_i - n\bar{x}\bar{y}}{\sum(x_i^2 - n\bar{x}^2) - \sum(y_i^2 - n\bar{y}^2)}$$

$$= \frac{\sum x_i y_i - n\bar{x}\bar{y}}{(\sum x_i^2 - \frac{(\sum x_i)^2}{n})(\sum y_i^2 - \frac{(\sum y_i)^2}{n})}$$

One can use any of the formulae given above.

Example: Following table gives the data of weight and diameter of tablets. Find out whether there is any relationship between weight and diameter of tablets.

Weight of tablets X (gm)	10.4	10.8	11.1	10.2	10.3	10.2	10.7	10.5	10.8	11.2	10.6	11.4
Diameter of tablets Y (mm)	7.4	7.6	7.9	7.2	7.4	7.1	7.4	7.2	7.8	7.7	7.8	8.3

Solution: To observe if there is any relationship, plot weight and diameter of tablets

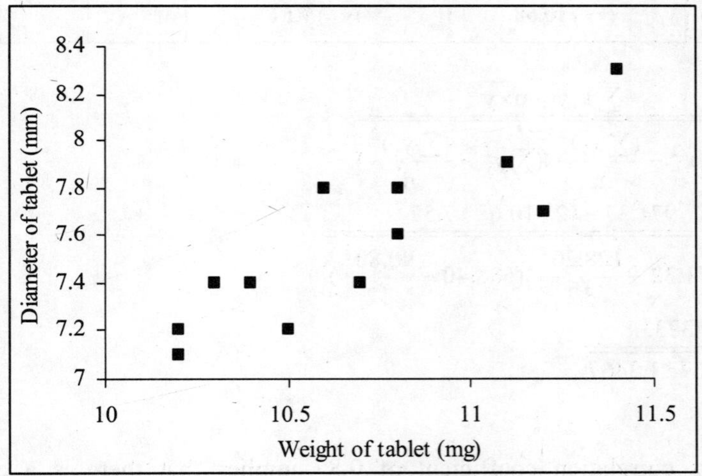

Figure: Scatter plot for weight and diameter of tablets

Scatter plot reveals that weight increases with diameter. Hence there is a relation between weight and diameter. Also as this relation is linear, we calculate Pearson's correlation coefficient as follows:

Table: Required calculations for Pearson's correlation coefficient

Tablet No.	Weight of tablets X (gm)	Diameter of tablets Y (mm)	X^2	Y^2	XY
1	10.4	7.4	108.16	54.76	76.96
2	10.8	7.6	116.64	57.76	82.08
3	11.1	7.9	123.21	62.41	87.69
4	10.2	7.2	104.04	51.84	73.44
5	10.3	7.4	106.09	54.76	76.22

6	10.2	7.1	104.04	50.41	72.42
7	10.7	7.4	114.49	54.76	79.18
8	10.5	7.2	110.25	51.84	75.6
9	10.8	7.8	116.64	60.84	84.24
10	11.2	7.7	125.44	59.29	86.24
11	10.6	7.8	112.36	60.84	82.68
12	11.4	8.3	129.96	68.89	94.62
Total	128.20	90.8	1371.32	688.40	971.37
Mean	$(\bar{x} =)\ 10.68$	$(\bar{y} =)\ 7.57$			

$$r = \frac{\sum x_i y_i - n\bar{x}\bar{y}}{\sqrt{\left(\sum x_i^2 - \frac{(\sum x_i)^2}{n}\right)\left(\sum y_i^2 - \frac{(\sum y_i)^2}{n}\right)}}$$

$$= \frac{971.37 - 12*10.68*7.57}{\sqrt{\left(1371.32 - \frac{128.20^2}{12}\right)\left(688.40 - \frac{90.80^2}{12}\right)}}$$

$$= \frac{1.3233}{\sqrt{1.7167*1.3467}}$$

$$= 0.87$$

Pearson's correlation coefficient of 0.87 implies that there is a correlation between weight and diameter of tablets.

Example: Following is the examination score obtained in mathematics and biology by 10 students. Find out if there is any relationship between the examination score of mathematics and biology.

Student Number	1	2	3	4	5	6	7	8	9	10
Mathematics Examination Score	57	45	72	78	53	63	86	98	59	71
Biology Examination Score	83	37	41	84	56	85	77	87	70	59

Solution: First plot examination score in Mathematics and examination score in Biology to observe if there is any relationship.

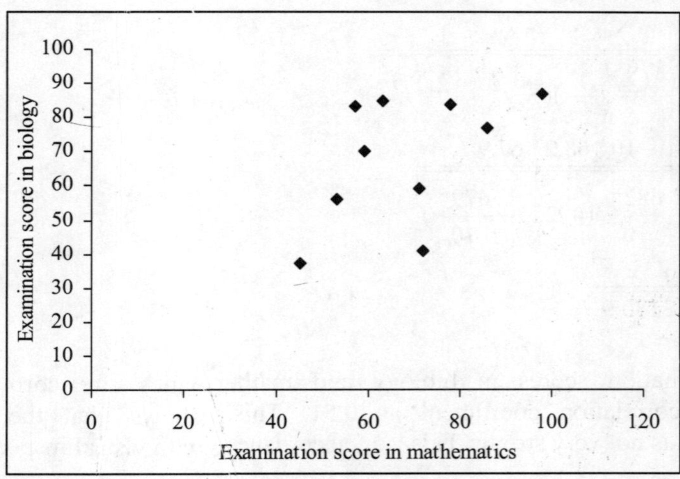

Figure: Scatter plot for examination score in mathematics and examination score in biology

Scatter plot reveals that examination score in biology and mathematics increases together except for few data points. Hence visually there is a relation between examination scores in biology and mathematics, though not a very good one. As this relation seems linear, we calculate Pearson's correlation coefficient as follows:

Table: Required calculations for Pearson's correlation coefficient

Student No.	Examination Score in Biology (X)	Examination Score in Mathematics (Y)	X^2	Y^2	XY
1	57	83	3249	6889	4731
2	45	37	2025	1369	1665
3	72	41	5184	1681	2952
4	78	84	6084	7056	6552
5	53	56	2809	3136	2968
6	63	85	3969	7225	5355
7	86	77	7396	5929	6622
8	98	87	9604	7569	8526
9	59	70	3481	4900	4130
10	71	59	5041	3481	4189
Sum	682	679	48842	49235	47690
Mean	$(\bar{x} =)$ 68.2	$(\bar{y} =)$ 67.9			

$$r = \frac{\sum x_i y_i - n\overline{x}\,\overline{y}}{\sqrt{\left(\sum x_i^2 - \frac{(\sum x_i)^2}{n}\right)\left(\sum y_i^2 - \frac{(\sum y_i)^2}{n}\right)}}$$

$$= \frac{47690 - 10*68.2*67.9}{\sqrt{\left(48842 - \frac{682^2}{10}\right)\left(49235 - \frac{679^2}{10}\right)}}$$

$$= \frac{1382.2}{\sqrt{2329.6*3130.9}}$$

$$= 0.51$$

The examination scores in biology and mathematics are correlated with Pearson's correlation coefficient as 0.51. This shows that the extent of relationship is not very strong. This is in accordance with visual inspection.

Spearman's Rank Correlation

This correlation coefficient is used when data is given as ranks. This is a non parametric measure of correlation. It is denoted by r_s.

$$r_s = 1 - \frac{6\sum_{i=1}^{n} d_i^2}{n^3 - n}$$

where d_i is the deviation of ranks for each experimental unit and n is the number of units.

Example: If instead of marks, we give grades to the examination scores in biology and mathematics (of above example), those will be as follows:

Student No.	Mathematics Examination Score(x)	Grade	Biology Examination Score(x)	Grade
1	57	3	83	7
2	45	1	37	1
3	72	7	41	2
4	78	8	84	8
5	53	2	56	3
6	63	5	85	9
7	86	9	77	6
8	98	10	87	10
9	59	4	70	5
10	71	6	59	4

Find out if there is any relationship between the grades of mathematics and biology.

Solution: To see the relationship between Mathematics and Biology grades, we plot the scatter diagram as follows:

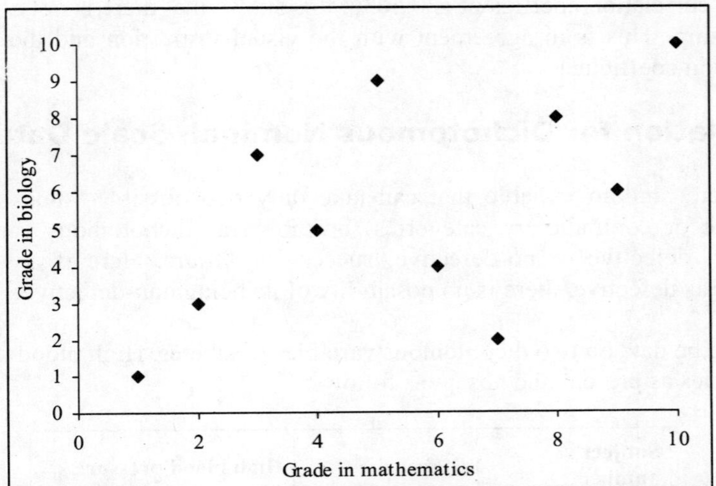

Figure: Scatter plot for grades of mathematics and grades of biology

This plot reveals that there is a relationship but not very strong. Let us now calculate the rank correlation and see the extent of relationship.

Table: Required calculations for rank correlation

Student	Grade in mathematics (x_i)	Grade in biology (y_i)	$d_i = x_i - y_i$	d_i^2
1	3	7	-4	16
2	1	1	0	0
3	7	2	5	25
4	8	8	0	0
5	2	3	-1	1
6	5	9	-4	16
7	9	6	3	9
8	10	10	0	0
9	4	5	-1	1
10	6	4	2	4
			$\sum d_i^2$	72

$$r_s = 1 - \frac{6\sum\limits_{i=1}^{n} d_i^2}{n^3 - n} = 1 - \frac{6*72}{10^3 - 10} = 1 - 0.4364 = 0.5636$$

As rank correlation coefficient is 0.5636, this shows that there is not a very good relationship. This is in agreement with the visual inspection and the Pearson's correlation coefficient.

Correlation for Dichotomous Nominal–Scale Data

A discrete random variable that can take only two possible values (mutually exclusive or contradictory categories) is known as dichotomous variable, for example, defective or no-defective, success or failure. Here if a variable is detected as defective, there is no possibility of its being non-defective.

Let there be data on two dichotomous variables (Diabetes, High blood pressure) with values as present and absent as follows:

Subject's number	Diabetes	High blood pressure
1	Present	Absent
2	Absent	Present
3	Absent	Absent
4	Present	Present
5	Absent	Absent
6	Present	Present

To calculate correlation coefficient for this case, first data is presented in a 2x2 table as follows:

		1st variable		
		Present	**Absent**	
2nd variable	**Present**	f_{11}	f_{21}	$R_1 = f_{11} + f_{21}$
	Absent	f_{12}	f_{22}	$R_2 = f_{12} + f_{22}$
Total		$C_1 = f_{11} + f_{12}$	$C_2 = f_{21} + f_{22}$	$f_{11} + f_{12} + f_{21} + f_{22}$

For the above mentioned data of diabetes, high blood pressure, this 2x2 table will be

		Diabetes		
		Present	**Absent**	
High blood pressure	**Present**	2	1	3
	Absent	1	2	3
Total		3	3	6

To measure the extent of correlation for this situation, many methods have been suggested.

The basic problem with the correlation measures of dichotomous variables is that there are two completely opposite situations.
When $f_{12} = f_{21} = 0$; this is the 'perfect correlation' situation. Both the variables are in perfect sync.
When $f_{11} = f_{22} = 0$; this corresponds to the 'worst correlation' situation. In this case if one variable is present, other will not.

Thus, the necessary conditions for a good measure of association is
  • Distinguish between these two opposite situations
  • Show a gradual and consistent change in the measure value when the values of variables change from the worst to the best case scenario.

We describe few measures of association: φ_2 Coefficient, Yule's Coefficient and Ives & Gibbons coefficient as follows:

φ_2 Coefficient

$$\varphi_2 = \frac{f_{11}f_{22} - f_{12}f_{21}}{\sqrt{C_1 C_2 R_1 R_2}}$$

The coefficient φ_2 satisfies both the above mentioned conditions. However, the φ_2 coefficient may not be defined if any of the factors in the denominator of the equation becomes zero.

Yule's Coefficient

$$Q = \frac{f_{11}f_{22} - f_{12}f_{21}}{f_{11}f_{22} + f_{12}f_{21}}$$

It is a good statistical measure of the association. It also satisfies the conditions for a good measure. Q is not defined when either f_{12} or $f_{21} = 0$ and either f_{11} or $f_{22} = 0$.

Ives and Gibbon's Coefficient

$$\varphi_2 = \frac{(f_{11} + f_{22}) - (f_{12} + f_{21})}{(f_{11} + f_{22}) + (f_{12} + f_{21})}$$

This measure also has good characteristics. It is always defined and it satisfies both the required conditions.

Example: Following table gives the presence/absence of a disease and presence/absence of an insect for a sample of 14 plants.

Plant No.	Presence of plant disease	Presence of insect
1	+	+
2	+	+
3	-	-
4	-	+
5	+	+
6	-	+
7	-	-
8	+	+
9	-	+
10	-	-
11	+	+
12	-	-
13	+	+
14	-	+

Find out if there is any association between disease of plant and insect in the plant.

Solution:

Table: 2x2 table of the data

		Plant Disease		
		Present	Absent	Total
Insect	**Present**	6	4	10
	Absent	0	4	4
	Total	6	8	14

$$\varphi_2 = \frac{f_{11}f_{22} - f_{12}f_{21}}{\sqrt{C_1 C_2 R_1 R_2}} = \frac{6*4 - 4*0}{\sqrt{6*8*10*4}} = 0.55$$

Yule's Coefficient: $\varphi_2 = \dfrac{f_{11}f_{22} - f_{12}f_{21}}{f_{11}f_{22} + f_{12}f_{21}} = \dfrac{6*4 - 4*0}{6*4 + 4*0} = 1.00$

Ives & Gibbons: $\varphi_2 = \dfrac{(f_{11}+f_{22})-(f_{12}+f_{21})}{(f_{11}+f_{22})+(f_{12}+f_{21})} = \dfrac{(6+4)-(4+0)}{(6+4)+(4+0)} = 0.43$

As f_{21} is 0 in this example, Yule's Coefficient should not be method of choice. We can observe that other two methods do not give very different values.

Concordance Correlation

If the objective of collecting pairs of data is to assess reproducibility of measurements, an effective technique is concordance correlation (Lin, 1989). It is denoted by r_n and is given by

$$r_n = \frac{2\sum (x_i - \bar{x})*(y_i - \bar{y})}{\sum (x_i - \bar{x})^2 + \sum (y_i - \bar{y})^2 + (n-1)(\bar{x} - \bar{y})^2}$$

Example: To test the reproducibility of analyses of lead concentrations in brain tissue using 2 different spectrophotometers, an experiment was conducted. Eleven tissue samples were taken and lead concentrations were measured using two spectrophotometers (Spectrophotometer 1 and Spectrophotometer 2). Results are as follows:

Tissue sample	Lead concentration in tissue (µg/g) by Spectrophotometer 1 (X_i)	Lead concentration in tissue (µg/g) by Spectrophotometer 2 (Y_i)
1	0.22	0.21
2	0.26	0.23
3	0.30	0.27
4	0.33	0.27
5	0.36	0.31
6	0.39	0.33
7	0.41	0.37
8	0.44	0.38
9	0.47	0.40
10	0.51	0.43
11	0.55	0.47

Find out whether both the spectrophotometers give similar results.

Solution: To calculate concordance correlation, first we calculate the various terms as in the following table:

Table: Required calculations for concordance correlation

Tissue sample	X_i	$X_i - \overline{X}$	$(X_i - \overline{X})^2$	Y_i	$Y_i - \overline{Y}$	$(Y_i - \overline{Y})^2$	$(X_i - \overline{X})*(Y_i - \overline{Y})$
1	0.2200	-0.1655	0.0274	0.2100	-0.1236	0.0153	0.0205
2	0.2600	-0.1255	0.0157	0.2300	-0.1036	0.0107	0.0130
3	0.3000	-0.0855	0.0073	0.2700	-0.0636	0.0040	0.0054
4	0.3300	-0.0555	0.0031	0.2700	-0.0636	0.0040	0.0035
5	0.3600	-0.0255	0.0006	0.3100	-0.0236	0.0006	0.0006
6	0.3900	0.0045	0.0000	0.3300	-0.0036	0.0000	0.0000
7	0.4100	0.0245	0.0006	0.3700	0.0364	0.0013	0.0009
8	0.4400	0.0545	0.0030	0.3800	0.0464	0.0021	0.0025
9	0.4700	0.0845	0.0071	0.4000	0.0664	0.0044	0.0056
10	0.5100	0.1245	0.0155	0.4300	0.0964	0.0093	0.0120
11	0.5500	0.1645	0.0271	0.4700	0.1364	0.0186	0.0224
Total	0.3855		0.1075	0.3336		0.0705	0.0865

$$r_n = \frac{2\sum(x_i - \overline{x})*(y_i - \overline{y})}{\sum(x_i - \overline{x})^2 + \sum(y_i - \overline{y})^2 + (n-1)(\overline{x} - \overline{y})^2}$$

$$r_n = \frac{2*0.0865}{0.1075 + 0.0705 + 10*(0.3855 - 0.3336)^2} = \frac{0.173}{0.2049} = 0.8445$$

As concordance correlation is 0.8445, the two spectrophotometers produce similar analyses of lead concentration in brain tissue. Hence both the spectrophotometers are suitable for analysis.

Intraclass Correlation

Sometimes it is desired to study the correlation between some characteristic of members of one or more families. Intra class correlation is used for this purpose. It is denoted by r_I and given by:

$$r_I = \frac{GroupMS - ErrorMS}{GroupMS + ErrorMS}$$

Example: Weight of twins is given in the following table:

Group (i.e. Twins)	Weight of one member of the twin group (kg)	Weight of other member of twin group (kg)
1	70.4	71.3
2	68.2	67.4
3	77.3	75.2
4	61.2	66.7
5	72.3	74.2
6	74.1	72.9
7	71.1	69.5

Find out if there is any correlation between weights of twins.

Solution:

Table: ANOVA table for the data

Source of variation	SS	DF	MS
Group	198.3143	6	33.0524
Member	0.4829	1	0.4829
Error	21.37714	6	3.5629
Total	220.17	13	

$$r_I = \frac{GroupMS - ErrorMS}{GroupMS + ErrorMS} = \frac{33.0524 - 3.5629}{33.0524 + 3.5629} = 0.8054$$

As r_I is 0.81, it implies that there is a good correlation between weights of twins.

Be cautious when there is non linear relationship

Suppose we have two variables X and Y as follows:

X	Y
-2	0
-1	3
0	4
1	3
2	0

If we calculate Pearson's correlation coefficient, it is 0 which implies that **there is no correlation**. On plotting this data, we get

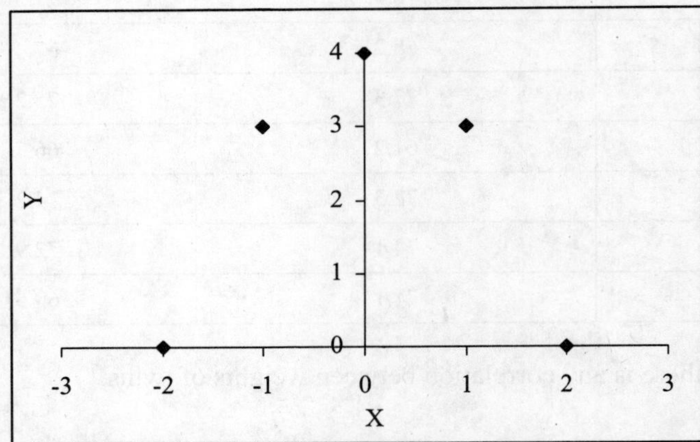

Figure: Scatter plot for two variables X and Y

This graph implies that there is a relationship between X and Y but not linear. Actually, there is a non linear relationship that is $Y = 4 - X^2$.

Similarly consider another example

X	Y
-3	9
-2	4
-1	1
1	1
2	4
3	9

Here again if we calculate Pearson's correlation coefficient, it is 0 which implies that **there is no correlation**. However, on plotting this data, we get

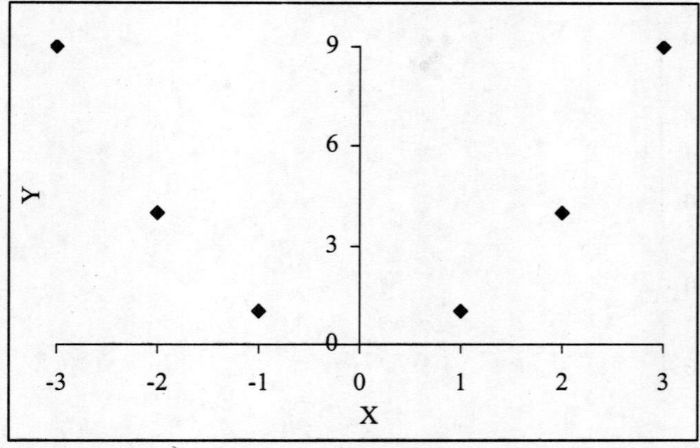

Figure: Scatter plot for two variables X and Y

This graph implies that there is a relationship between X and Y but not linear. Actually, there is a non linear relationship that is $Y=X^2$.

Be careful when calculating Pearson's correlation coefficient as it is appropriate only for linear relationships.

Hence before calculating Pearson's correlation coefficient, look at the scatter diagram, if relationship appears linear, then only calculate Pearson's correlation coefficient.

Be cautious when data is with extreme values

Now let us consider a situation as follows.

X	Y
2.0	48
2.1	50
1.9	50
2.0	46
2.1	47
5.5	12

Here Pearson's correlation coefficient r is -0.994 which implies that there is a very good correlation but if we calculate rank correlation coefficient r_s, it is -0.1714.

On plotting this data, we get

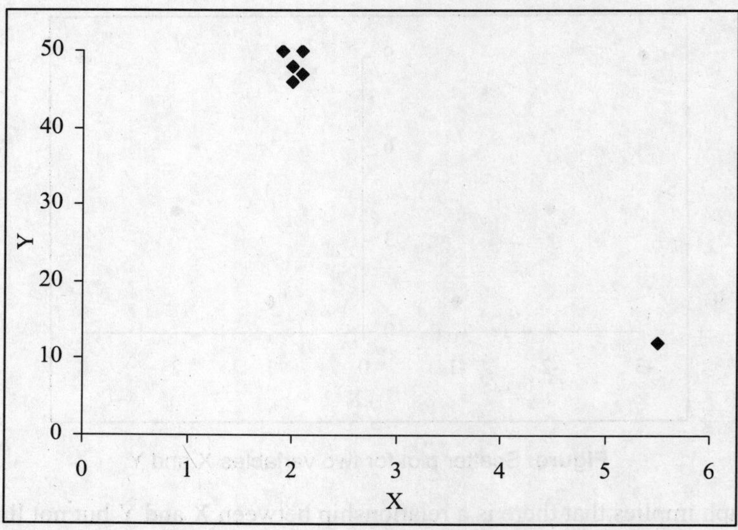

Figure: Scatter plot for two variables X and Y

This plot shows that data is clustered around the value 47 and one data (12) is very far away from this. We remove this one data point of (5.5, 12) and plot again.

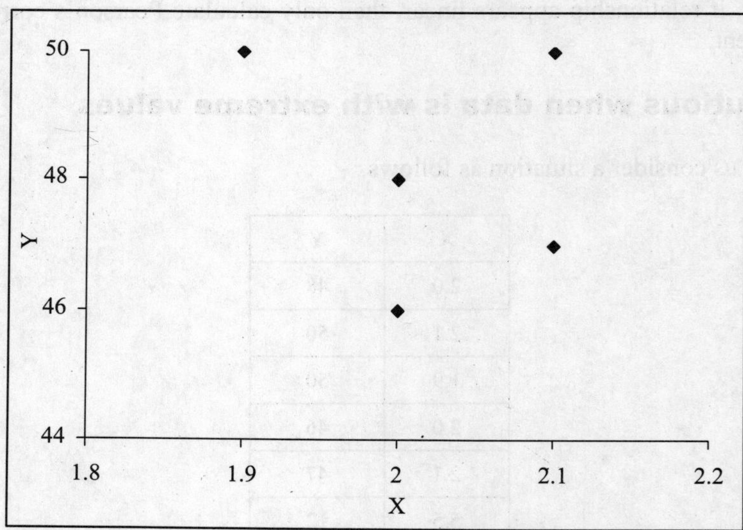

Figure: Scatter plot for two variables X and Y after deleting the data point (5.5, 12)

We see that though values of X increase, but there is not much change in Y values, i.e., there does not seem any relationship between two variables. Moreover, Pearson's correlation coefficient r is -0.200 after removing the data point (5.5, 12) which is in line with the above calculated rank correlation coefficient.

This example implies that when there are extreme values, we should calculate rank correlation coefficient rather than Pearson's correlation coefficient.

SIMPLE LINEAR REGRESSION

Once we know that there is a correlation between two variables, our interest is in knowing the mathematical form of relationship between them.

Ascertaining the probable (mathematical) form of the relationship between variables comes under regression analysis.

The objective for establishing a relationship between two variables is to know how the magnitude of one of the variables (the dependent variable) is a function of the magnitude of the second variable (the independent variable) and consequently, predict the dependent variable based on independent variable.

Dependent variable: Dependent variable is the one which is "Predicted"
Independent variable: Independent variable is the one which is "Predictor"

Always Remember
If one variable is a function of other variable, it dose not mean that other variable is also a function of first variable. For example,

- Weight is a function of age but vice versa is not true.
- Clinical response is a function of dose administered but other way it may not be correct.

So the first step is to check which one is the dependent variable and which one is the independent variable.

Ultimate objective in regression analysis is to estimate the value of dependent variable corresponding to a given value of independent variable as accurate as possible.

The simplest functional relationship of one variable to another is the simple linear relationship (i.e., line) as follows:

$$Y_i = \alpha + \beta X_i$$

Now to estimate dependent variable Y using the independent variable X, we need to know coefficients α and β. We may have many lines for various values of α and β. When we predict Y based on X, there is always some difference between actual value of Y and predicted value of Y, i.e., there is an error associated with predicted value of Y, so equation becomes

$$Y_i = \alpha + \beta X_i + \varepsilon_i$$

where ε_i denotes the error in predicted value of Y_i.

To understand this let us consider the example below:

Example: Following table gives the dissolution data of a drug.

Dissolution time (min) X	% Dissolved Y
3.0	14
4.0	15
5.0	22
6.0	28
8.0	39
9.0	42
10.0	50
11.0	55
12.0	57
14.0	60
15.0	65
16.0	72
17.0	74

Solution: % Dissolved as a function of dissolution time is plotted.

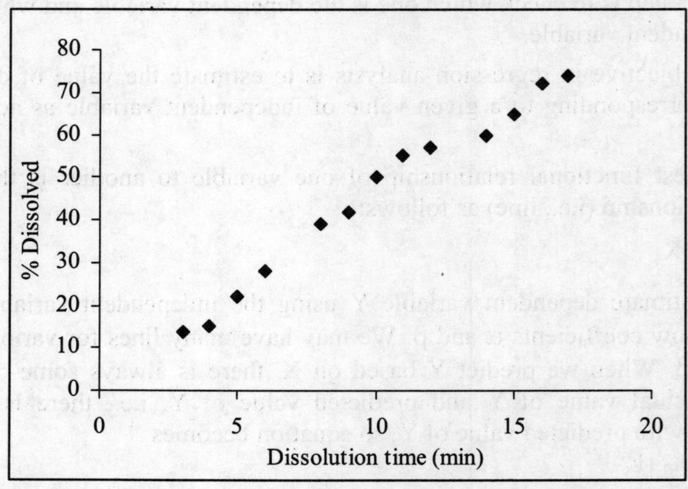

Figure: Scatter plot of dissolution time vs. % dissolved

If we wish to plot lines through these pairs of data, many lines can be drawn as follows:

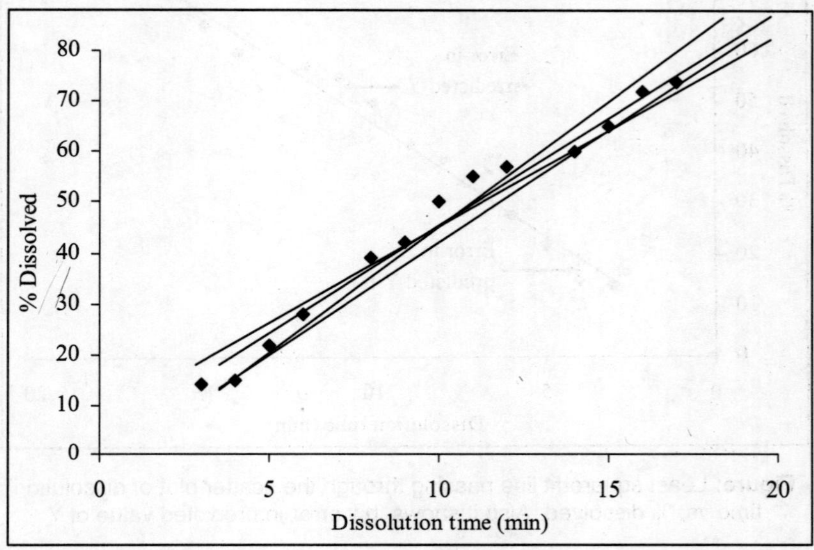

Figure: Various lines passing through the scatter plot of dissolution time vs. % dissolved

The question is now which line should be used for predicting Y. To get the accurate estimate of Y, we wish to fit the best line through these points. So we will apply least square method to get the best fit line which will give the estimates "a" and "b" of α and β. The equation will become:

$$\hat{Y}_i = a + bX_i$$

where $\hat{Y}_i$ is the predicted value of Y_i.

Concept of least squares

This is based on minimizing the sum of squares of errors ($\sum_i \varepsilon_i^2$) in predicting the dependent variable.

These errors are deviations of actual value (Y_i) and predicted value ($\hat{Y}_i$) of dependent variables.

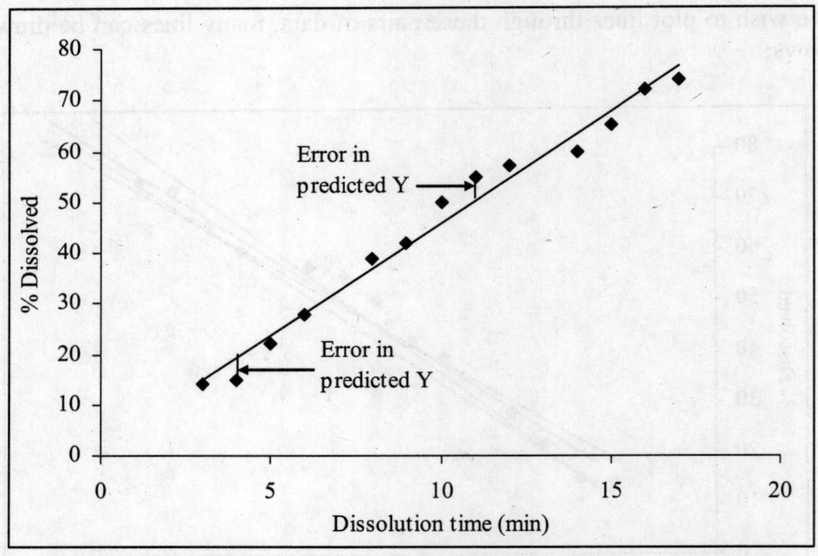

Figure: Least square fit line passing through the scatter plot of dissolution time vs. % dissolved. Also it shows thw error in predicted value of Y

This is nothing but the sum of squares of the vertical distances of actual point from the fitted line (refer the arrows in the above figure).

This **sum of squares of errors** is denoted by $\sum (Y_i - \widehat{Y}_i)^2$.

$\sum (Y_i - \widehat{Y}_i)^2$ is also known as residual sum of squares.

So we have to find "a" and "b" such that **sum of squares of errors** is minimum.

On applying optimization techniques, we get "a" and "b" as follows:

$$b = \frac{\sum\limits_{i=1}^{n}(x_i - \bar{x})(y_i - \bar{y})}{\sum\limits_{i=1}^{n}(x_i - \bar{x})^2} = \frac{\sum x_i y_i - n\bar{x}\bar{y}}{\left(\sum x_i^2 - \dfrac{(\sum x_i)^2}{n}\right)}$$

$$a = \bar{y} - b\bar{x}$$

So now Y can be estimated as $\widehat{Y}_i$ using the following equation:

$$\widehat{Y}_i = a + bX_i$$

When we apply regression analysis, following assumptions should be met:
- The sample must be representative of the population for the inference prediction.
- The independent (predictor) variables are error-free.
- The error is assumed to be a random variable with a mean of zero.
- The variance of the error is constant across observations.
- The errors are uuncorrelated.

Remember that regression equation should be used to estimate Y only for the range for which the equation is developed and it should not be used to estimate the value of Y out of that range

Following are few common scenarios in pharmaceutical science where regression is used. One should be very careful in interpreting the results of those relationships as those will be applicable only for the range for which those relationships are built up.

- The analytical method should be used only for the range for which it is developed.
- Dose response relationship should predict the response only in the range for which the equation is developed.
- Pharmacokinetic – pharmacodynamic relationship should infer about the relationship only in the dose range for which it is developed.
- Pharmacokinetic investigation of dose proportionality should be considered only for the doses studied.

Example: Stability data with triplicate assays from a single batch is given as follows.

Time (weeks)	0	8	16	24	32
Concentration (% of label)	102, 102, 104	100, 99, 101	98, 99, 98	94, 97, 96	97, 95, 93

Find out if there is any relationship between time and concentration data.

Solution: Calculate the average concentration at each time point and then plot time on x axis and average concentration on Y axis.

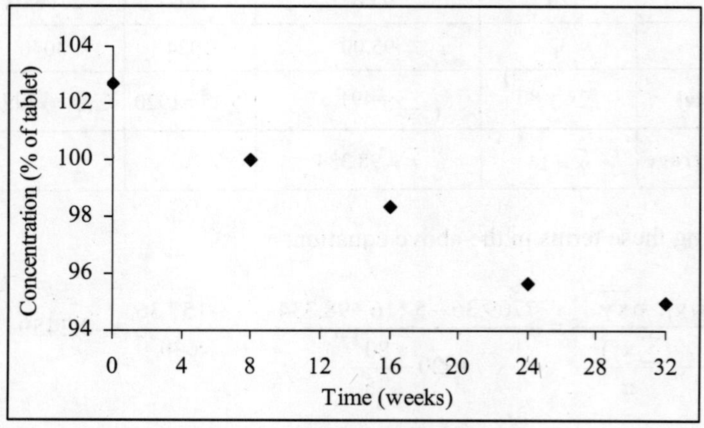

Figure: Scatter plot of concentration vs. time

Graph infers that Concentration (% of label) decreases as time increases i.e. there is a negative correlation. As the relationship is linear, Pearson's correlation coefficient (r) will ¨give the extent of relationship. Pearson's correlation coefficient is -0.9861 which shows a very good correlation between assay concentration and weeks.

So the next step is to establish a mathematical form of this relationship. As there is linear relationship, we can fit the following equation:

$$\hat{Y}_i = a + bX_i$$

where

$$b = \frac{\sum_{i=1}^{n}(x_i - \bar{x})(y_i - \bar{y})}{\sum_{i=1}^{n}(x_i - \bar{x})^2} = \frac{\sum x_i y_i - n\bar{x}\bar{y}}{\left(\sum x_i^2 - \frac{(\sum x_i)^2}{n}\right)}$$

$$a = \bar{y} - b\bar{x}$$

Let us first calculate the various terms in the formula.

Table: Required calculations for regression analysis

	Time (weeks) (x)	Mean concentration (y)	x^2	xy
	0	102.67	0	0
	8	100.00	64	800
	16	98.33	256	1573.28
	24	95.67	576	2296.08
	32	95.00	1024	3040
Total	$\sum x = 80$	$\sum y = 491.67$	$\sum x^2 = 1920$	$\sum xy = 7709.36$
Average	$\bar{x} = 16$	$\bar{y} = 98.334$		

Substituting these terms in the above equation:

$$b = \frac{\sum x_i y_i - n\bar{x}\bar{y}}{\left(\sum x_i^2 - \frac{(\sum x_i)^2}{n}\right)} = \frac{7709.36 - 5*16*98.334}{1920 - \frac{80^2}{5}} = = \frac{-157.36}{640} = -0.2459$$

$$a = \bar{y} - b\bar{x} = 98.334 - (-0.2459)*16 = 102.268$$

Hence the least square equation is

$$\hat{Y}_i = 102.268 - 0.2459 X_i$$

$\Rightarrow$ Assay concentration = 102.268-0.2459 * Time(in weeks)

This equation can be used for predicting the assay concentration at various time points. Now let us calculate the assay concentrations at 0, 8, 16, 24, and 32 weeks.

Table: Observed concentrations and predicted concentrations using regression line

Time (weeks)	Concentration (% of label)	Mean observed concentration	Predicted concentrations obtained using regression equation
0	102, 102, 104	102.67	102.27
8	100, 99, 101	100.00	100.30
16	98, 99, 98	98.33	98.33
24	94, 97, 96	95.67	96.37
32	97, 95, 93	95.00	94.40

The table values reveal that predicted assay concentrations are very close to the experimental assay concentrations. Hence the equation can be used for prediction purposes very accurately.

Example: An experiment was conducted to determine composition of binary mixture of water and glycerine by viscosity measurement using Ostwald's U tube viscometer. This experiment requires preparation of a series of standard binary mixtures of water and glycerine and then measuring the viscosity. Observations are as follows:

Sr. No.	1	2	3	4	5
% of glycerine (x)	2	4	6	8	10
Viscosity in centipoises (y)	1	1.8	3	4	4.9

Find out the relationship between viscosity and % of glycerine content.

Solution: First plot % of glycerine vs. viscosity to observe if there is any relationship.

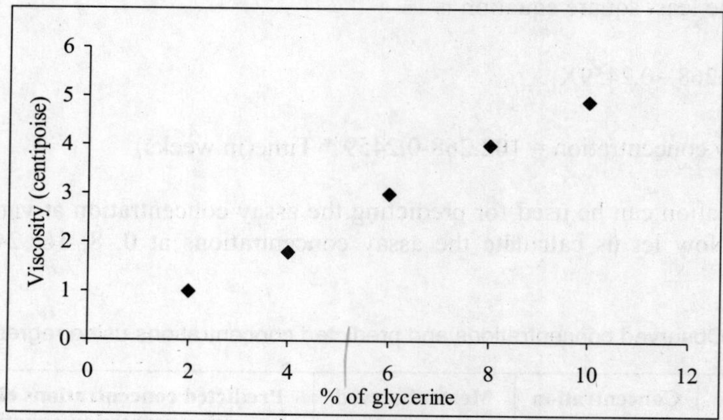

Figure: Scatter plot of viscosity vs. % of glycerine

Scatter plot reveals that viscosity increases proportionally with % of glycerine. Hence there is a good relationship between viscosity and % of glycerine. Also this relationship is linear. So we calculate Pearson's correlation coefficient as follows:

Table: Required calculations for correlation and regression analysis

Preparation No.	% of glycerine (X)	Viscosity (Y)	X^2	Y^2	XY
1	2	1	4	1	2
2	4	1.8	16	3.24	7.2
3	6	3	36	9	18
4	8	4	64	16	32
5	10	4.9	100	24.01	49
Sum	30	14.7	220	53.25	108.2
Mean	6	2.94	44	10.65	21.64

$$r = \frac{\sum x_i y_i - n\overline{x}\,\overline{y}}{\sqrt{(\sum x_i^2 - \frac{(\sum x_i)^2}{n})(\sum y_i^2 - \frac{(\sum y_i)^2}{n})}}$$

$$= \frac{108.2 - 5*6*2.94}{\sqrt{(220 - \frac{30^2}{5})(53.25 - \frac{14.7^2}{15})}}$$

$$= \frac{20}{\sqrt{40 * 10.032}} = \frac{20}{20.03}$$

$$= 0.9984$$

Viscosity and % of glycerine are strongly correlated as Pearson's correlation coefficient is 0.9984.

Now the next step is to establish a mathematical form of this relationship. As there is a linear relationship, we can fit the following equation:

$$\hat{Y}_i = a + bX_i$$

where

$$b = \frac{\sum_{i=1}^{n}(x_i - \bar{x})(y_i - \bar{y})}{\sum_{i=1}^{n}(x_i - \bar{x})^2} = \frac{\sum x_i y_i - n\bar{x}\bar{y}}{(\sum x_i^2 - \frac{(\sum x_i)^2}{n})}$$

$$a = \bar{y} - b\bar{x}$$

Substituting the various terms in the above equation:

$$b = \frac{\sum x_i y_i - n\bar{x}\bar{y}}{(\sum x_i^2 - \frac{(\sum x_i)^2}{n})} = \frac{108.2 - 5*6*2.94}{(220 - \frac{30^2}{5})} = \frac{20}{40} = 0.5$$

$$a = \bar{y} - b\bar{x} = 2.94 - (0.5)*6 = -0.06$$

Hence the least square equation will be

$$\hat{Y}_i = -0.06 + 0.5X_i$$

$$\Rightarrow \text{Viscosity} = -0.06 + 0.5 * \% \text{ of glycerine}$$

This equation can be used for predicting the viscosity at various % of glycerine. Now let us calculate the viscosity at various % of glycerine.

Table: Observed viscosity and predicted viscosity using regression line

Preparation No.	% of glycerine (X)	Observed viscosity (Y)	Viscosity obtained using regression equation ($\hat{Y}$)
1	2	1	0.94
2	4	1.8	1.94
3	6	3	2.94
4	8	4	3.94
5	10	4.9	4.94

The table values reveal that predicted viscosity is very close to experimental viscosity. Hence the equation can be used for prediction purposes with accuracy.

Example: Standard curve of isoniazid was prepared in the concentration range of 0.2 µg/ml to 1.25 µg/ml. Standard curve data is given below:

S.No.	Concentration (µg/ml)	Absorbance
1	0.1	0.010
2	0.25	0.021
3	0.5	0.051
4	0.75	0.065
5	1	0.090
6	1.25	0.120

Plot the standard curve and find out the correlation coefficient and linear regression.

Solution:
Following figure gives the standard curve plot.

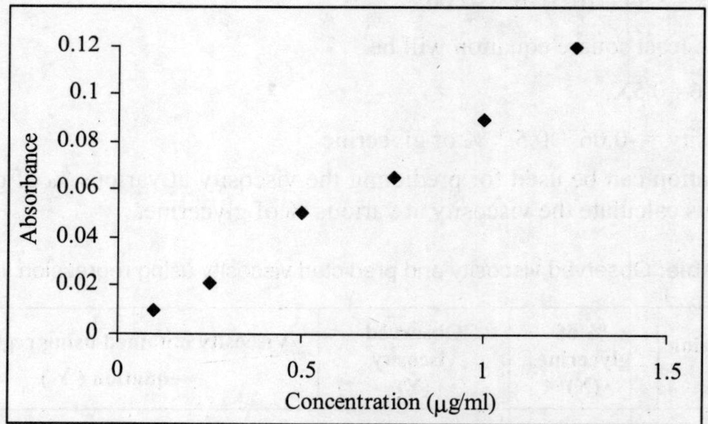

Figure: Standard curve plot of absorbance vs. concentrations

This plot infers that there is a linear relationship between concentrations and absorbance. Also the correlation coefficient is 0.9918. This confirms a strong linear relationship between concentration and absorbance.

So using the least square method, we get the regression equation as follows:
Absorbance = 0.0936 * Concentration -0.0005

This equation can be used for predicting the concentrations at various values of absorbance.

Hence this equation was used to calculate the amount of isoniazid released from the pellets, 10 mg, at the end of 30 min in dissolution test, where the dissolution media used was water (900ml) and apparatus used was USP type I. The dissolution results for six units (i.e., absorbance for 6 units) are given in the following table. The amount of drug released was calculated using this equation and results are as follows:

Sr. No.	Absorbance	Amount of drug released at 30 min (D.F= 100, 900 ml) (mg)	% drug released at 30 min
1	0.082	7.932692	79.33
2	0.091	8.798077	87.98
3	0.085	8.221154	82.21
4	0.087	8.413462	84.13
5	0.095	9.182692	91.83
6	0.097	9.375	93.75

Absorbance = 0.0936 * Concentration -0.0005

Concentration = (Absorbance + 0.0005)/ 0.0936

If absorbance =0.082

Concentration = (0.082+0.0005) / 0.0936 = 0.88141 µg / 100 ml

So amount released at the end of 30 min in a volume of 900 ml = 0.88141 X 900 / 100 = 7.932692 mg

% Drug released at the end of 30 min = (Calculated amount / Initial amt) * 100

$$= (7.932692 / 10) * 100 = 79.33 \%$$

Example: Application of regression in acute toxicity: Defining acute toxicity based only on the numeric value of an LD_{50} is not right. The relationship between dose and response is more important in risk assessment than the numeric value of LD_{50}. A slope can give a clue to the mechanism of toxicity.

Multiple regression and correlation

If there are simultaneous measurements for more than 2 variables, and one of the variables is dependent on others, then we are dealing with multiple regression and correlation. As computation in this case is complex, we will not discuss here.

To summarize

- Correlation is concerned with measuring the strength of the relationship between variables
- Regression establishes the relationship between dependent and independent variables

I can prove anything by statistics except the truth.

— George Canning

Chapter 13

Outlier Detection

WHAT IS AN OUTLIER

- An outlier is an observation that lies outside the overall pattern of a distribution (Moore and McCabe 1999).
- That means statistically, outliers refer to relatively small or large values which are considered to be different from, and not belong to, the main body of the data. Outliers are often easy to spot in histograms.

QUESTIONS TO BE ASKED

In fact, the presence of such observations may be a clue to an important process inherent in the experimental system. Hence following issues should be checked before declaring the observation(s) as outlier(s):

- Was the outlier due to an error of observation?
- Can we expect such extreme values to occur routinely in such experiments?
- Do the outlying data really represent the experimental process that is being investigated?

STRATEGIES WHICH CAN BE FOLLOWED

If answer to first question is yes,
- If possible, repeat the appropriate portion of the experiment,
    - Estimate the missing values
    - Analyze the data without discarded values
- Otherwise analyze the data after discarding those values

If answer to first question is no and answer to other questions is also no, we conclude that the data point is an outlier and has come from a population different from the one studied.

HOW TO DETECT OUTLIER

Only looking at the data, if an observation(s) is declared as outlier(s), it is a subjective decision i.e., one person may say it is outlier other may not.

So we need some objective method to detect outlier(s). Statistical methods are used for this purpose.

However, these methods should be used if no obvious cause for the outlier can be found.

STATISTICAL METHODS TO DETECT OUTLIERS

Several statistical methods are available to detect the outlying observations.

First arrange the observations according to the magnitude. If some observation(s) seems to be outlier visually, those observation(s) can be tested for outlier(s).

The T procedure

Calculate T_n statistic as follows:

$$T_n = \frac{x_n - \bar{x}}{sd}$$

where x_n is the extreme observation (it can be on lower side or upper side), $\bar{x}$ is the mean of the data and sd is the standard deviation of the data.

If the calculated value of T_n is more than the value obtained from "T procedure" table (tabulated value) at α level of significance, declare the extreme observation, x_n, as outlier.

Example: Following table gives the cholesterol level of 15 subjects.

Subject	Cholesterol
1	165
2	188
3	194
4	197
5	200
6	202
7	205
8	210

9	214
10	215
11	227
12	231
13	239
14	249
15	297

Solution: Here data is already arranged in increasing order. Let us assume that subject no. 1 and 15 are outliers. We will perform T procedure one by one.

$$T_n = \frac{x_n - \bar{x}}{sd} = \frac{165 - 215.5}{30.9} = -1.63$$

As Tabulated value T_n for N=15 is 2.549, subject no 1 (165) is not an outlier.

$$T_n = \frac{x_n - \bar{x}}{sd} = \frac{297 - 215.5}{30.9} = 2.64$$

As 2.64 is greater than 2.549, subject no 15 (297) can be considered as an outlier.

Dixon's test for extreme values

Calculate the ratio (r) of the difference of suspected value from one of its nearest neighboring values to the range of the observations. The formula for the ratio r, depends on the sample size as shown in the following table.

Table: Formulae for r statistic for various sample size

Smaple size	If suspected value is the largest value	If suspected value is the smallest value
3 to 7	$r = \dfrac{x_n - x_{n-1}}{x_n - x_1}$	$r = \dfrac{x_2 - x_1}{x_n - x_1}$
8 to 10	$r = \dfrac{x_n - x_{n-1}}{x_n - x_2}$	$r = \dfrac{x_2 - x_1}{x_{n-1} - x_1}$
11 to 13	$r = \dfrac{x_n - x_{n-2}}{x_n - x_2}$	$r = \dfrac{x_3 - x_1}{x_{n-1} - x_1}$
14 to 25	$r = \dfrac{x_n - x_{n-2}}{x_n - x_3}$	$r = \dfrac{x_3 - x_1}{x_{n-2} - x_1}$

The **null hypothesis** associated to Dixon's test is as follows: "There is no significant difference between the suspect value and the rest of them, any differences must be exclusively attributed to random errors".

Thus, for testing the largest value (x_n) or the smallest value (x_1) as possible outliers, compare this calculated value of r to "r value" obtained from the Dixon's test of outlier tables (tabulated value). If calculated value is more than tabulated value, the suspected value can be characterized as an outlier and it can be discarded during analysis, if not, the suspected value must be retained and used in all subsequent calculations. To check for outliers other than extreme values, the Dixon test can be repeated, however, the power of this test decreases as the number of repetitions increases.

Example: Data from the above example

Subject	Cholesterol
1	165
2	188
3	194
4	197
5	200
6	202
7	205
8	210
9	214
10	215
11	227
12	231
13	239
14	249
15	297

For a sample size of 15, formula to test the lowest value as an outlier from the above table is

$$r = \frac{x_3 - x_1}{x_{n-2} - x_1} = \frac{194 - 165}{239 - 165} = \frac{29}{74} = 0.39$$

As "r value" obtained from the Dixon's test of outlier table is 0.525 for n =15 and tabulated value or "r" is less than this, 165 can not be considered as an outlier.

For a sample size of 15, formula to test the highest value as an outlier from the above table is

$$r = \frac{x_n - x_{n-2}}{x_n - x_3} = \frac{297 - 239}{297 - 194} = \frac{58}{103} = 0.56$$

As "r value" obtained from the Dixon's test of outlier table is 0.525 for n =15 and tabulated value or "r" is more than this, 297 can be considered as an outlier.

Winsorizing

- In this method, the extreme values, both low and high, are changed to the values of their closest neighbors.
- This method is also useful when there are missing values and those are known to be the extreme values.

Always remember that statistics can be a very useful tool to make decisions, but it should not be used as only yardstick.

To summarize
- In general aberrant observations should not be arbitrarily discarded only because they look too small or too large
- One should be very cautious while declaring any observation(s) as outlier because those are not easily accepted and always looked upon with suspicion

I'm not an outlier; I just haven't found my distribution yet!

— **Ronan Conroy, Ireland**